Critical Lands Protected with Sansom's Help

Brazos River

Colorado River

Devils River

DALLAS

FORT WORTH

AUSTIN

SAN ANTONIO

HOUSTON

GULF OF MEXICO

CELEBRATING

50 YEARS

Texas A&M University Press
publishing since 1974

ANDREW SANSOM
A Life in Conservation

Kathie and Ed Cox Jr. Books on Conservation Leadership

SPONSORED BY

THE MEADOWS CENTER
FOR WATER AND THE ENVIRONMENT

TEXAS STATE UNIVERSITY

Andrew Sansom, General Editor

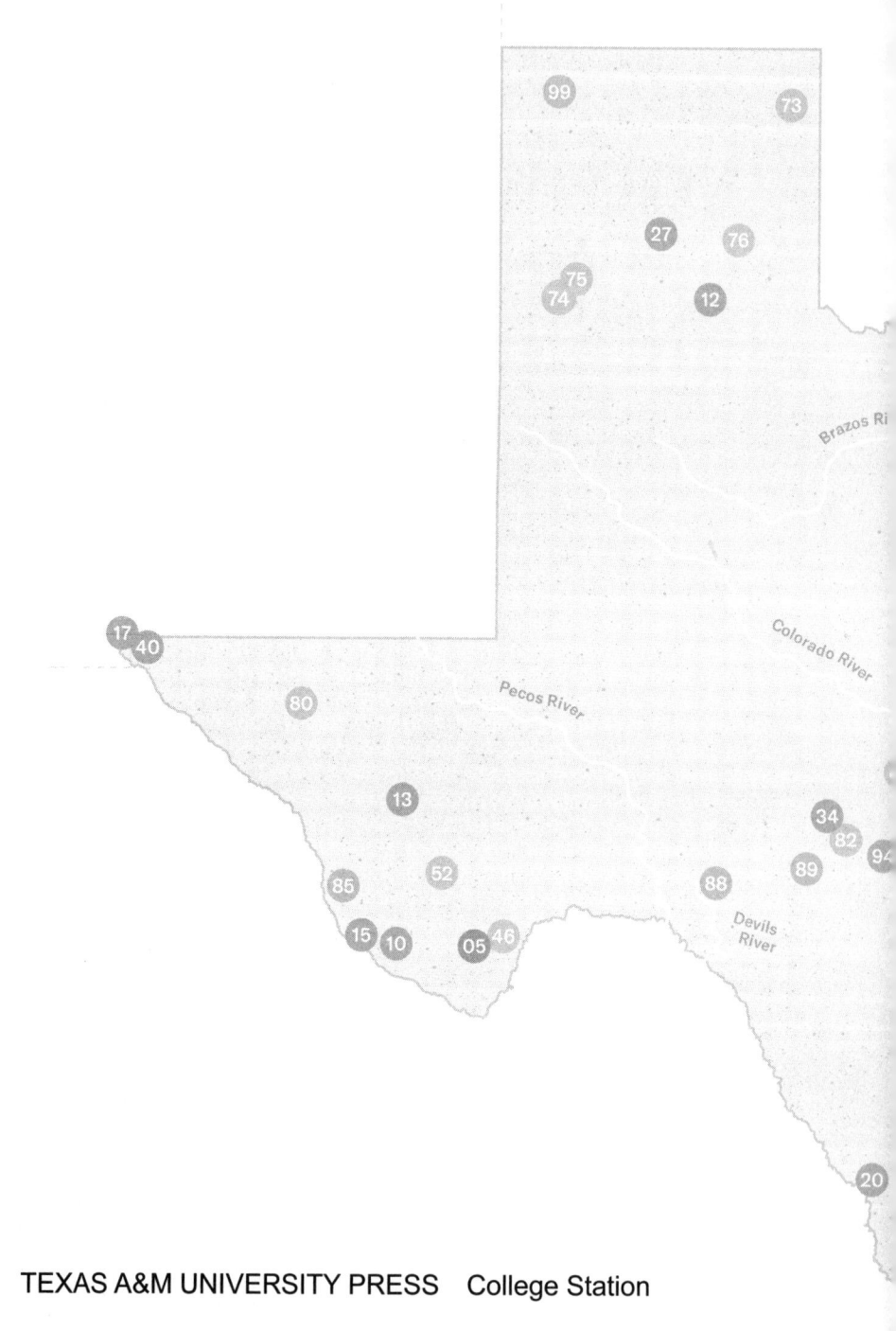

TEXAS A&M UNIVERSITY PRESS   College Station

Andrew Sansom

A Life in Conservation

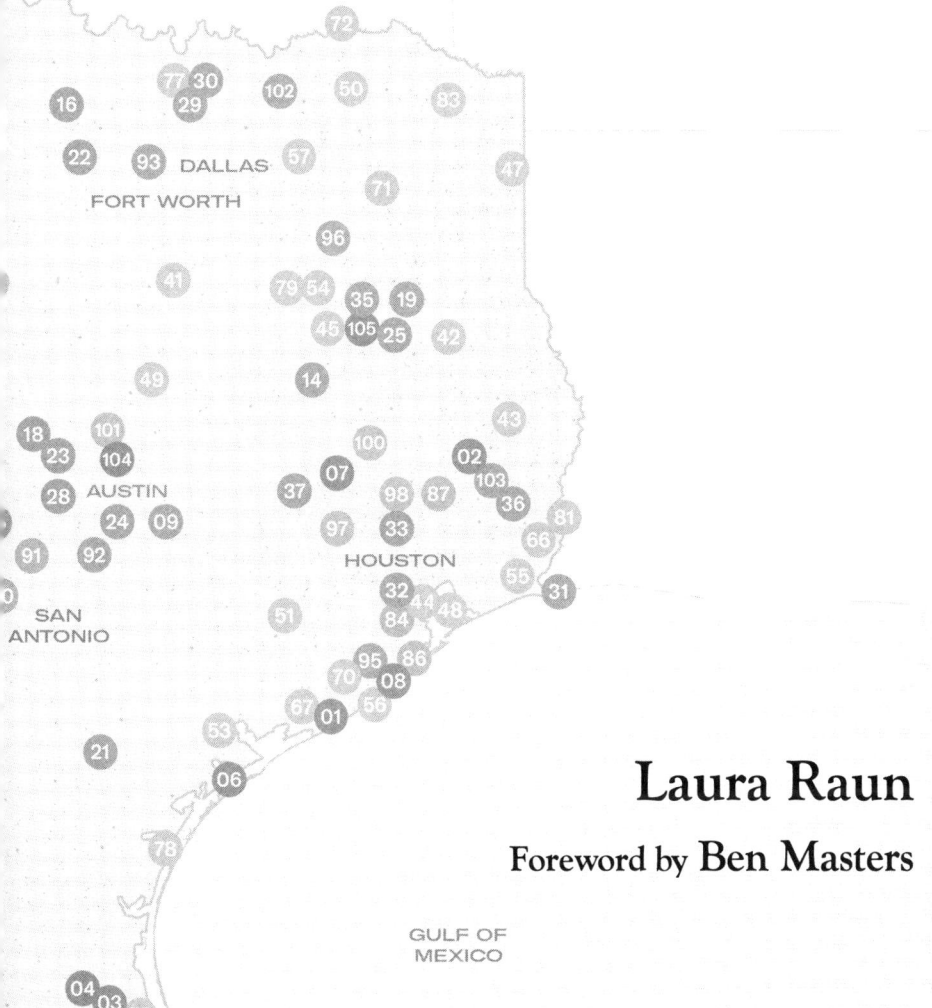

Laura Raun

Foreword by Ben Masters

∞ This paper meets the requirements of ANSI/NISO Z39.48–1992
(Permanence of Paper).
Binding materials have been chosen for durability.

Library of Congress Cataloging-in-Publication Data

Names: Raun, Laura, author.
Title: Andrew Sansom: a life in conservation / Laura Raun.
Other titles: Kathie and Ed Cox Jr. books on conservation leadership.
Description: First edition. | College Station: Texas A&M University Press,
[2024] | Series: Kathie and Ed Cox Jr. books on conservation leadership
| Includes bibliographical references and index.
Identifiers: LCCN 2024022202 (print) | LCCN 2024022203 (ebook) |
ISBN 9781648432460 (cloth) | ISBN 9781648432477 (ebook)
Subjects: LCSH: Sansom, Andrew. | Water conservation—Texas |
Conservationists—Texas—Biography. | Natural
Resources—Texas—Management. | Environmental management—Texas. |
Conservation of natural resources—Political aspects—Texas. |
Environmentalism—Political aspects—Texas. | BISAC: BIOGRAPHY &
AUTOBIOGRAPHY / Environmentalists & Naturalists | BIOGRAPHY &
AUTOBIOGRAPHY / Personal Memoirs
Classification: LCC TD224.T4 R38 2024 (print) | LCC TD224.T4 (ebook) |
DDC 333.95/16092—dc23/eng/20240709
LC record available at https://lccn.loc.gov/2024022202
LC ebook record available at https://lccn.loc.gov/2024022203

A list of titles in this series is available at the end of the book.

To my husband Sam Carroll,
who introduced me to Andy in the first place.

The conservation of natural resources is the fundamental problem. Unless we solve that problem it will avail us little to solve all others.

—Theodore Roosevelt

Contents

Foreword

Andy Sansom has many great qualities that have led him to become one of the most important conservationists in Texas history. His loyalty to his beloved home state is unmatched, his passion for landscape conservation is contagious, and his leadership of the Texas Parks and Wildlife Department led to the conservation of more than one hundred parks and protected areas. But my favorite quality of Andy Sansom by far is a trait that only springs to the surface when his boots are off the payment and his binoculars are in hands—his childlike enthusiasm when watching wildlife.

I first met Andy at a stuffy fundraiser in San Antonio. He was in his coat and tie shaking hands, shaking wallets, and preaching the gospel of conservation to business leaders in Alamo Heights. I introduced myself and told him I was a Texas-based wildlife filmmaker. He asked me if I had ever visited the largest bat cave in the world, Bracken Cave. I told him no, I had not been there, but it was on my list. He placed his hand on my shoulder, stared me in the eyes, and said that it was absolutely imperative that I go, and soon, before the bats migrated south for the winter. Before I could respond, he pulled Nona aside, checked his calendar, and told me that we would go to Bracken Cave together the following week because the bats needed a film to help their cause.

We met at the Bracken Cave parking lot a few nights later. Andy had traded his suit and tie for old blue jeans and well-worn boots. He had the energy of a college freshman attending their first football game. As we walked to the cave, he pointed out a Swainson's Hawk circling in the sky above, getting ready for the night's big event. We walked through oaks and junipers to a depression in the ground with a thirty-foot cave mouth at the far side. A few bats were fluttering at the cave entrance. Andy started to get excited, and we followed him to a bench to wait for their emergence.

Rat snake! Andy pointed towards some cactus above the cave entrance, and sure enough, a four-foot rat snake was getting into position. Peregrine Falcon! Andy squinted skyward at a tiny speck in the sky. Coachwhip! Andy gestured at a reddish-tan snake making its way through the rocks. He looked at me with glee and smiled when he told me they're the fastest snake in the United States . . . and Texas too.

The bats began to emerge as the sun set. At first it was a few dozen, then a few hundred, and then the entire cave entrance was swirling as 15 million bats took flight to hunt insects for the night. I looked over at Andy, and he was smiling from ear-to-ear hole. He pointed to the sky and yelled out "here comes the Peregrine!" and swoosh the falcon stooped and grabbed a bat. "Quick, watch the coachwhip" as the snake plunged through the rocks to nab a scrambling bat. "Look at those Swainsons, they're hunting as a group!" He started pumping his fist and cheered when the bat got away. He grabbed my arm and started pointing at the rat snake, poised on a cactus. "Wait for it . . . steady. . . . Now!" he exclaimed and fist pumped as the snake struck through the air to grab a flying bat. Nobody roots for the home team like Andy Sansom.

Andy Sansom cheers for nature louder than the Aggies yell at Kyle Field. He gets more excited than when the Red Raiders upset the Sooners, or when the Rangers win the World Series. He's on team wildlife, and team wild landscapes, and he's brought countless wins to his home turf of Texas. Of course, his "wins" are much more important than a game, a season, a trophy, or a title. His conservation wins are the greatest legacy that a generation can leave behind—Wild Places, where nature can thrive, and Texans can fill our souls with crisp air and clean water and have room to roam. Like thousands of other Texans, I've been inspired by Andy Sansom's love for wildlife in a way that transcends maps and protected areas and easements and inflows. Andy instilled in me a love of Texas and a patriotic sense of duty to serve the natural wonders of our magnificent state. Years after our evening at Bracken Cave, Andy encouraged me to produce and direct Texas' first ever wildlife movie, which I named *Deep in the Heart: A Texas Wildlife Story*, after the love we mutually share for our home state. And yes, there is a bat sequence at Bracken Cave inspired by Andy Sansom.

This book does far more than chronicle a man's life, it tells the story of opportunities that were presented to a generation and seized by a bold leader who was looking multiple generations into the future. Texans hundreds of years from now will walk down trails conserved by Andy Sansom. They'll marvel at the bald cypress, be awed by the state bison herd, and cheer the bats that escape the hawks. Thank you, Andy, for doing your damndest to leave Texas a better place. And thank you, Laura, for documenting this critical part of our state's history.

—Ben Masters

Preface

Andy Sansom has a cult following, and several years ago some of his devotees wanted him to write a memoir to divulge the secrets behind his success as an environmental conservationist. They recognized that Sansom was the most accomplished protector of critical lands in Texas in recent memory, or perhaps ever, based on acreage and geographical spread. Andy said he didn't have time for a memoir, much less an autobiography.

Talks ensued between Andy, Shannon Davies, and me about some kind of book that would chronicle his life. A biography is what Andy chose, and it seemed like the best way of discovering what one admirer called his "secret sauce." The mystery was how Sansom was able to help protect a half-million acres of land for the public good in a state like Texas that prized private property. Natural resources in the Lone Star State are more typically viewed as assets to monetize than as vital ecosystems to sustain for people, plants, and animals.

Leadership was a key ingredient in Sansom's secret sauce, and his style differed diametrically from most in the 2010s and 2020s, when authoritarianism gained currency. Many young people don't know a different kind of leadership other than one driven by narcissism, rigidity, and ideology. Many older people have forgotten what civilized governance looks like. Sansom's leadership was grounded in collaboration, pragmatism, and humility—and based on evidence.

His ability to chart new paths, sometimes radical ones, in environmental protection while working inside the system provides valuable lessons for conservationists, public policy makers, elected officials, and voters. I believe Sansom has valuable lessons for a world that is at an environmental tipping point.

I first met Sansom in 2003 and worked with him from time to time as a public relations practitioner specializing in the water industry. I had set up my PR firm in 2001 after a career as a journalist, working for the *Financial Times*, CNN, Bloomberg News, and other media companies.

As a former business and financial journalist, I had interviewed and written about scores of industrial captains, including Freddy Heineken and Steve Ball. I knew that a book would be a much more comprehensive effort, so I fell back on journalistic instincts of outlines, probing questions, and colorful stories.

What I didn't know was how different book writing would be from journalistic writing. What an epiphany! I soon learned that every paragraph would take three times longer than in a news story—more research, more thought, and more insight. It took four years.

My goal was twofold. One was to discover how Andy was able to protect a half-million acres of land for public use in a state where land for private use dictates most political, economic, and social priorities. My second goal was to examine Andy's life to see how he responded to the universal challenges facing all human beings. I believe our unique reactions to the same human condition are what distinguish us, one from another. The book in your hands is my attempt to achieve these goals.

Acknowledgments

When Andy chose to have a biography written about him, rather than writing a memoir or autobiography himself, I took that to mean he was willing to expose his life to public scrutiny. He did not balk, and for his good graces, I am profoundly grateful.

That decision of Andy's also exposed his family—wife Nona, son Andrew, and daughter April, in particular. To them, I am equally grateful.

From the outset, the discussions about biography versus autobiography versus memoir included Shannon Davies, former director of Texas A&M University Press. Shannon is a world-class editor, and I have worked with some of the best in New York and London. She is a remarkable book coach who was willing to guide me from the world of journalistic writing to the world of book writing.

Also involved from the outset of the project was David Todd of the Conservation History Association of Texas, who provided financial and moral support throughout. A good number of Andy's admirers contributed to the cause as well. They included J. David Bamberger, Ramona and Lee Bass, the Chiltepin Charitable Fund (in honor of Shannon Davies), Carol E. Dinkins, the Conservation History Association of Texas, Will and Pam Harte, the Harte Charitable Foundation, Karen Hixon, Ben Jones, The Meadows Foundation, the Cynthia & George Mitchell Foundation, Nancy and Ted Paup, Pam Reese, the Shield-Ayers Foundation, Eric Sumberg, Sheila and Walter Umphrey, the Way Family Foundation, and the Winkler Family Foundation.

Instrumental to this book was Roberta Coffin, who helped with everything from basic research, to checking land transaction details, to curating content for the map of lands protected with Andy's help. Erin West, the most gifted graphic designer I've ever worked with, created the map, capturing Andy's legacy in one masterful image.

I owe a debt of gratitude to Loren Steffy and Wendy Gordon for reviewing my manuscript and providing invaluable feedback that saved me from the most egregious embarrassment, though probably not all.

I also am indebted to Marguerite Avery, senior acquisitions editor at TAMU Press, for guiding me through the book process with a firm hand on the tiller.

One of the most challenging aspects of chronicling Andy's life was reconstructing the early years of environmental conservation in Texas. Ken Kramer proved invaluable insight in that regard, as did Jim Blackburn.

Andy's role in various water issues was sketched in helpful detail by Warren Pulich and Cindy Loeffler.

Karl Rove recounted political and personal journeys that he and Andy shared, while John Howard masterfully recollected Andy's time in the administration of Governor George Bush.

Robert Mace provided his brilliant, witty, and incisive take on anything I asked him. Molly Stevens offered a treasure trove of background on environmental education in Texas.

And finally, to anyone else who should have been mentioned here—thank you.

ANDREW SANSOM

Chapter 1

You're Gonna Need a New Ear

With any luck, Andrew Sansom would have made it to Lake Jackson on time to speak to the Brazosport Birders. But in a flash, the oncoming car suddenly turned in front of him.

With no time to swerve, he hit the woman's vehicle head-on, and everything went black. More than a few minutes went by before the whirring sound of helicopter blades entered his consciousness. The Life Flight chopper had landed at the scene near West Columbia, about sixty miles south of Houston. Andy's head was being cradled in the lap of a Good Samaritan who had stopped at the crash site when the emergency responder arrived from the helicopter. Lying on the pavement nearby was Andy's right ear.

"Well," said the emergency responder, "we're gonna have to get you a new ear."

Sansom had another idea. "Put it on ice and let's see if they can put it back on."

The next two weeks in May of 2005 were filled with a couple of days of intensive care and a series of surgeries at Memorial Hermann Hospital in Houston, a pillar of the Texas Medical Center, itself the largest medical complex in the world. Head and face wounds were mended by plastic surgeons, while his right leg was stitched together after being broken in two places below the knee. Support poured in from friendships built up over a lifetime.

One concerned friend showed up in person. The morning after the accident Sansom looked up and at the foot of his bed saw Nolan Ryan—the baseball legend who had perfected the fastball, been inducted into the Hall of Fame, and run the Texas Rangers major league team. Incidentally, Ryan was also a former commissioner at the Texas Parks and Wildlife Department when Sansom ran the agency.

What caught the attention of the hospital staff even more was a phone call that came soon after Sansom arrived. US Senator Kay Bailey Hutchison wanted to know how Sansom was doing. The next thing he knew, he was moved "upstairs" to one of the private suites usually reserved for Saudi princes and other royalty.

Yet another friend demanded to know, "Where's the car?" Walter Umphrey also had been a Parks and Wildlife commissioner during Sansom's tenure from 1990 to 2001 and was a prominent plaintiff's lawyer, having won billions of dollars in verdicts and settlements from asbestos litigation. Sansom's wife, Nona, had called Umphrey to explain why she and Andy wouldn't make dinner on Friday night.

Umphrey, who had also helped the state of Texas negotiate a historic $17.3 billion settlement with tobacco companies, sent an accident investigator to inspect Sansom's car the next day. It was an early model Toyota Prius, which had started selling in the United States only a couple of years before. The investigator told Nona shortly thereafter that the car had saved Andy's life and that he would be dead if the car hadn't done everything it was designed to do to protect the driver. Meanwhile, the driver of the other car in the accident was unhurt.

In the end, Sansom's severed ear wasn't put back on his head. A plastic surgeon came to Sansom's room and told him the best ears were available from M. D. Anderson, the renowned cancer hospital down the street. The ears were prosthetic, given the difficulty of making them from flesh. And they were less than perfect, according to one of Sansom's former commissioners at Texas Parks and Wildlife whose father had one. The ex-commissioner said that when her father got hot and sweaty, the ear would slide down his face. Sansom decided he didn't want a meltable ear.

And so the one-eared man became a kind of legend, at least in a small and remote village in Alaska. Sansom regularly fished the seventy-five-mile Kanektok River, known for abundant salmon and trout, which empties into the Bering Sea near the Inuit village of Quinhagak in southwest Alaska. One year, he and his party arrived in Quinhagak a day earlier than the float plane was scheduled to pick them up. They camped just outside the village and spent the day playing with the children, who were fascinated by the people from the Lower 48.

They were most fascinated by Sansom and his missing ear. "My loss of an ear is a child magnet," Sansom explained. The kids followed him around, laughed, made rabbit-ear Vs with their fingers, and took pictures.

A decade later, Sansom and his friends were back in Quinhagak after nearly a month on the river. He was standing in the checkout line at a local general store, and a young boy walked up to him.

"We heard about you," the boy said, pointing in awe at the right side of Andy's head.

Chapter 2

Growing up on the Riverbanks

He was born on a naval base located in an unlikely spot, the desert of Nevada. It was October 10, 1945, and his father had only recently returned from military service in the South Pacific. Sansom was the first baby born in the hospital at the Hawthorne Naval Ammunition Depot, as the facility then was known. He was named Andrew Heacock Sansom, his middle name being his mother's maiden name.

The elder Sansom, Ernest, was a chemist and munitions specialist during World War II and had grown up in Texas. Given his expertise as a chemist, Ernest was drawn to his native state as the petrochemical industry there had burgeoned during the war. Demand for synthetic chemicals and rubber had fueled the growth. Before Sansom was one year old, his parents moved back to Texas, and his father landed a job as a research chemist with the Dow Chemical Company in Lake Jackson on the Texas coast. The young family settled into a planned community built by Dow just a few years earlier for workers at the new plant. The 1,300-acre community was designed by Alden B. Dow, who was a son of the chemical company founder and who was heavily influenced by Frank Lloyd Wright. The residential development accounted for the bulk of the town, which also came to include Dow-designed schools, a church, a theater, and other buildings that featured midcentury modern architecture situated amid ample green spaces. The town's layout included as many trees as possible set among the winding streets.

Lake Jackson was a unique town on the Texas Gulf Coast, where oil refineries and chemical plants had sprung up after the Texas oil boom of the early twentieth century. Coastal culture of the 1940s and '50s was driven by petrochemical companies, not by architects and urban planners who were more influential in the northeastern part of the United States, such as Levittown, Pennsylvania. Park land in coastal communities often meant a few trees planted around a Little League baseball field. Lake Jackson looked different. Sansom was aware of that as a child and more so as his world expanded. "My earliest memories were of being in the park lands and around the original structures Alden Dow designed, " Sansom said.

While Lake Jackson was largely a planned community, it also was a company town dominated by the industrial operations of Dow Chemical. When a warning siren went off to signal a plant emergency, families held their breath until they heard that all was well.

Lake Jackson was idyllic for a boy growing up in the 1950s. Open areas abounded for unsupervised play. The town was located on the banks of the Brazos River—a slow, red, meandering ribbon of water that is the longest river in Texas. It cuts diagonally across the state from northwest to southeast, emptying into the Gulf of Mexico amid the ubiquitous petrochemical towers, tanks, and pipelines that dot the landscape near Houston. The Sansom house sat on a winding tributary of the river called Oyster Creek, near the coastal wetlands of Brazoria County. From early on, water played a critical role in Sansom's life.

"We all engaged with the creek in one way or the other," recalled Sansom. "It was just a part of our life. My father and I built a boat for me to use, and I spent every day after that on the creek."

Sansom's childhood epitomized the Norman Rockwell culture of the white middle class at the time. His mother, Imo, was an avid gardener, having grown up on a farm in Alabama. She was a devoted homemaker and admired schoolteacher. She taught second-grade classes and had earned a master's degree by going to night school at the University of Houston, driving more than an hour one-way from Lake Jackson to attend classes.

It was his maternal grandmother who first instilled Sansom's love of water, even before his father helped him build the boat. The young Sansom was "big enough to walk with her down to the lake," perhaps four or five years old. There in Alabama, where his mother had grown up, his grandmother taught him to fish, using a cane pole and worms they dug up from the chicken yard and kept slithering in a bucket. Together, the grandmother and grandson traipsed to the big pond, and Sansom was hooked. Fishing would remain a cherished hobby for the rest of his life.

The Norman Rockwell childhood included a family pet. "I had only one dog growing up and loved him," Sansom recounted. "But we came home from church one Sunday to find him on the living room floor eating the roast my mother had left out to cool for our Sunday lunch. That's the last I ever saw of him."

Hunting figured prominently in Sansom's teenage years. "The first money I ever earned as a clerk in the local sporting goods store in Lake Jackson I spent on a shotgun," he recollected. "I spent those early years in the marshes along the coast with one of my closest friends with whom I still hunt each year today. I experienced millions of waterfowl migrating down to the rice fields and marshes in the winter, thus the spectacle alone drew me into the field the first time."

Sansom's father nurtured his son's interest in boating, fishing, and hunting even though he himself wasn't an outdoor sportsman. He also encouraged the youth's interest in the natural environment. The elder Sansom was an amateur archeologist and liked to pick up arrowheads unearthed by the continual digging to lay oil and gas pipelines for nearby petrochemical plants. "He was curious and interested in all kinds of things," Sansom explained, for example, the sulfur domes along the Texas coast—bulges on the flat landscape that were filled with sulfur, petroleum, and salt. They rose as much as seventy-five feet above sea level, and the Sansom family would visit them on Sunday afternoon outings.

The nuclear family also included Sansom's sister, Jeanne, who was a year and a half younger. She, too, played on Oyster Creek and took up fly fishing later in life, as did Sansom. In the 1950s, the tightknit family spent most Sunday afternoons going for family drives, a ritual of many households at that time in Texas, along the Texas coast. Family values prevailed in the Sansom clan, fueled by a Calvinistic heritage. The elder Sansom was the first male in three generations of his family not ordained as a Presbyterian minister. Sansom's earliest memories of his paternal grandfather were of him preaching from the pulpit.

While the pulpit was his most vivid memory, prisoners were his most influential memory. His grandfather lived in the church manse in Huntsville, Texas, about ten blocks from the Texas State Penitentiary, sometimes called "Walls Unit" because of its red brick walls. One time when Sansom was visiting his grandparents, a recently released inmate still wearing prison whites knocked at the door. "My grandfather would give them enough money to get new clothes and a bus ride to wherever their home was," recalled Sansom. The kindness and compassion shown by his grandfather made a lasting impression on the young boy.

Andrew, or Andy as he was widely known, sang in the church choir and started piano lessons in the second grade. His parents were strict and not averse to paddling when needed to impose discipline. Consequently, he was mostly obedient with occasional bouts of rebelliousness.

In general, life did not offer a lot to rebel against. Sansom was an intelligent and healthy boy. His parents were nurturing and financially able to care for him. He lived in a garden-like community that reflected architect Alden Dow's view of architecture as "the manifestation of wholesome living." It was ideal for running, playing, bicycling, and fishing.

Sansom's Rockwellian boyhood included the Boy Scouts, an organization depicted in many of Rockwell's works. Sansom's first scout trip was to Camp Karankawa near West Columbia, where the fateful car accident would occur years later. The boys camped overnight in old, holey army tents that leaked in the heavy rain and left them drenched and miserable. Yet Sansom was even more drawn to the outdoors.

He visited his first state park on a Boy Scout trip. It was Garner State Park, in the Texas Hill Country, when he was eight or nine years old. It provided a feast for the eyes—high mesas, canyons, limestone cliffs, and streams. Sansom couldn't know then that he would someday run the agency overseeing the park. What he did know was that his love of nature was growing.

What Sansom also took away from the scout trip to the park was a lesson in character formation. He and several other troop members decided to climb a steep bluff, including one boy who was clearly overweight. In the 1950s and '60s, it was common for youngsters to make fun of those who were heavyset, and scouts were no different. "We treated him badly," Sansom admitted. When the boys set out for the climbing expedition, "We couldn't believe that this fat kid was coming with us," Sansom continued. The youths were about three-quarters of the way up the bluff when "the fat kid panicked. He got stuck, he couldn't go any further."

By this time, the scoutmaster had figured out where the boys were and was at the bottom of the cliff, looking up and scowling. "Basically, what he said was that it was our job to get the fat kid to the top," Sansom recounted. The boys got underneath him and pushed him up the rest of the way. "After that, there was no more making fun, that was over," Sansom remembered vividly. "The change in perspective was that we realized he was one of us. Do unto others as you would have them do unto you."

When the young Sansom wasn't scouting, fishing, or paddling Oyster Creek, he was at the neighborhood swimming pool. Lake Jackson, a Levittown on the Brazos, had a community pool that was only a block from the Sansom house. Sansom could walk there, and by the time he was twelve years old, he was volunteering to help teach swimming lessons. In high school, he got his first paying job, as a lifeguard and swimming teacher. The pool provided the first inkling that it was possible to make a living off his love for the outdoors.

Making money wasn't the only perk of the pool. There were cute girls, too. Sansom was tanned, taut, and towheaded. And the girls noticed. They were even willing to tag along with him to close the pool in the evenings before heading off for a movie.

"One night, I shut the pool down and went to backwash it and clean the filters. I got so interested in my date that I forgot to turn the pump off. The next morning, I get these hysterical calls from the city manager saying that I drained the water tower in that part of town because I'd forgotten to turn the valve off in the pool. All the water for that whole section of town was in the ditches. How I survived the job I have no idea."

As a student, Sansom was mediocre, with a touch of attention deficit disorder, by his own admission. While he wasn't in the top of his class scholastically, he was popular, social, and musically talented. He played the role of Will Parker in the musical *Oklahoma*, the first musical production ever performed at Brazosport High School. He was voted "Most Representative" of his senior class, requiring him to make a speech at graduation. The early experience in publicly addressing his classmates was a precursor to later life roles in which he often spoke publicly in front of large groups and became a sought-after presenter.

The idyllic boyhood eventually collided with reality, however. While Sansom was in high school, his father went back to school to get his law degree, realizing the value of such expertise in the chemical industry. Eventually, he became one of the first patent lawyers in Texas, though the price was high. He commuted to Texas Southern University in Houston at night, after working all day in Lake Jackson—about an hour's drive each way. The stress mounted, and his family could see it. "He developed a horrible bleeding ulcer," Sansom said. "It almost killed him."

The physical toll on his father was a memory that stayed with Sansom, though he noted that both of his parents went back to school to get advanced degrees after working in the professional world. "It was almost like it was expected. This is what you do."

Still, Sansom's teenage years provided a solid bridge from childhood to adulthood, enabling him to indulge his love for the natural environment while also revealing ways of making a living from it.

The question was whether he could navigate life outside of the cloistered existence in Lake Jackson.

Chapter 3
Hippie Sixties

The student movement that would help define the "hippie sixties" was formalized in 1960 when young activists founded Students for a Democratic Society (SDS) and the Student Nonviolent Coordinating Committee (SNCC). Both were national activist organizations that sought to engage college students as agents of change for economic justice, racial equality, and international peace. Their calls for acts of civil disobedience, antiwar protests, and antiauthoritarian demonstrations rocked the country's major institutions, no more so than on university and college campuses.

In late 1964, students at the University of California at Berkeley mobilized to protest the university's restrictions on political speech and assembly, prompting the arrest of around eight hundred students. By 1965, campus protesters across the country were increasingly focused on the Vietnam War as President Lyndon B. Johnson ramped up US troop presence and bombing there. Students burned draft cards and chanted, "Hey, hey LBJ, how many kids did you kill today?" For many, it was an exhilarating time to be young and in school—an era of jeans and tie-dyed shirts, long hair, rock music, psychedelic drugs, and casual sex.

For Sansom, however, the mid-1960s were a bit tamer. The biggest question for him was where to go to college. The top candidate was Austin College, a small, liberal arts school in Sherman, Texas, affiliated with the Presbyterian Church. Although his parents didn't pressure him, the religious connection was a draw for Sansom because of his family background with the church. In September 1964, he enrolled at Austin College and moved 350 miles north to Sherman, near the Texas border with Oklahoma.

After he had flown the nest, Sansom floundered in his newfound freedom. He had no clear idea of what he wanted to study, so Western civilization served the purpose. Sansom's biggest accomplishment during his two years at Austin College was meeting the young woman who eventually would become his life partner. His path to matrimony started even before classes began. Hazing of incoming freshmen was a revered tradition at the private school. Students would be roused out of their dormitory beds at four in the morning and made to wear beanies and do cheers.

The consolation was the head cheerleader—a striking coed named Nona Bishop Wood. Sansom suffered through the humiliation of beanie -wearing just to be able to gaze at her running down the hallway in the dorm. And eventually he caught her eye. After a fraternity party at Lake Texoma, just a few miles from campus, Andy and Nona ended up at a pizza joint, each with other dates. The two couples sat across from each other.

"I looked up and across the table was this rosy-complected, blonde-haired man with dazzling blue eyes," recalled Nona. "And he was grinning at me."

She tried to send encouraging signals since women in the 1960s generally did not call men they were attracted to, and she was thrilled when he called her. They started dating and quickly were going steady, with much in common. Nona had grown up in Houston, an hour from Lake Jackson, with a similar coastal climate. She was majoring in elementary education, just as his mother had done.

Beyond Sansom's azure eyes, it was his kindness and interest in other people that attracted her. One story in particular touched her. It was about a longtime pen pal of Sansom's, a Cuban boy named Tomás Delgado. Sansom was "so concerned about this young man in Cuba who wanted to get out" that he mounted a monumental effort to make it possible for Tomás to come to the United States, Nona recalled fondly. This is a good man, she thought, even though Tomás didn't emigrate at that time.

By his second year at Austin College, Sansom started feeling that his studies weren't leading him on a path to professional fulfillment. His curiosity was piqued when his younger sister was looking at a Texas Tech University catalog in her search for a college. He noticed the university's parks and recreation management program, one of the best in the country. It had not occurred to him that he could actually get a degree in something he had been doing every summer for ten years or so. Andy convinced Nona to spend spring break of 1966 driving with him to visit Texas Tech in Lubbock, 325 miles away. Sansom interviewed at the university and Nona at a local school district for a teaching job, as she was graduating that spring. They moved to Lubbock for the fall semester, Andy living in a school dormitory and Nona in an apartment.

By this time, the couple was planning their wedding for the summer of 1967. But as they talked, they wondered why they should wait for another half of a year—why not just get married over Christmas? Nona called her mother to float the idea and got a good response. And so it happened on December 27, 1966. They tied the knot at Bethany Methodist Church in Houston.

Academically, Sansom was lukewarm on his major of parks and recreation management. He taught bowling and refereed football games on the recreation side, but there was no fire in the belly for his major. That is, until Professor James W. (Bill) Kitchen entered the picture. Kitchen supervised independent study projects, and Sansom registered for one. The assignment seemed herculean—locate, photograph, and write the history of every railroad tunnel in Texas. Sansom felt overwhelmed until he learned there were only five in the state, including an abandoned one near Fredericksburg that would reappear down the road.

From there, Kitchen introduced the young student to the canyons of the Texas Panhandle—Caprock Canyons and Palo Duro Canyon, which are some of the most dramatic in the United States. Sansom began to realize that a career in outdoor recreation was broader than just teaching swimming. Sansom had an epiphany, which he attributed to his mentor, Kitchen—"It involved the outdoors writ large and the need to protect it." The revelation led to the next major step in Sansom's discernment of his calling, and later in life, to his method of graduate teaching through independent projects.

In December of 1968, Sansom graduated from Texas Tech with a bachelor's degree in parks and recreation management. This put him in the bull's-eye of the military draft, which had been expanded as the Vietnam War raged on. The Military Selective Service Act of 1967 ended student deferments after completion of a four-year degree or your twenty-fourth birthday, whichever came first. Sansom was twenty-two. And the act expanded potential conscription to all males between eighteen and fifty-five. In addition, there was growing talk of instituting a lottery to select draftees.

Sansom did what many other twenty-two-year-old males did at that time—he enrolled in graduate school, though it was no guarantee against being drafted. Like other young men at that time, Sansom figured it was worth staying in school—in his case, studying parks and recreation—because the law might change, or the war might end.

In the spring of 1969, he landed a plum assignment as student chair of the organizing committee for the Southwest Park and Recreation Training Institute, a high-profile conference about parks and recreation that Texas Tech had hosted since 1956. Chaired by Sansom's mentor, Bill Kitchen, the conference brought together federal agency representatives who shared information about programs created in the early 1960s, such as the Bureau of Outdoor Recreation, the Land and Water Conservation Fund, and the Housing and Urban Development Act.

The students planned the speakers and logistics for the event, which was held at Texoma Lodge on Lake Texoma, where Andy and Nona had first caught each other's eyes. The conference keynote speaker was Bill Pond, the first director of the newly created National Recreation and Park Association, based in Washington, DC. The association had been formed in 1965 as a merger of five organizations: the National Recreation Association, the American Institute of Park Executives, the American Recreation Society, the National Conference on State Parks, and the American Association of Zoological Parks and Aquariums.

Pond flew into Love Field in Dallas, the nearest airport to the lodge in Mead, Oklahoma, and Sansom made sure he was the one to pick him up, drive him to the conference, and serve as his personal aide. During the chauffeuring, the ambitious Sansom recognized an opportunity to address two looming challenges—the risk of being drafted and the direction of his career.

Sansom chatted up Pond and managed to land a job offer as the first intern at the National Recreation and Park Association in Washington. An internship in the nation's capital could benefit Sansom in a couple of ways. It could bolster his claim to be a student, though that decision resided with the local draft board. It also offered an opportunity to explore some nascent thoughts of running for public office, including Congress.

The draft, however, threatened to derail those plans. In the spring of 1969, the Vietnam War was escalating, with deployments and casualties climbing. Sansom tried to join the National Guard in the Washington, DC, area and explored other military assignments to avoid being drafted, though all efforts failed.

Shortly thereafter, a draft notice arrived and Sansom accepted that he would have to serve if unable to get a deferment. While he didn't consider himself a "radical," he decided to appeal to his local draft board in Angleton, near his hometown of Lake Jackson. Using his innate skills of persuasion, he argued that he should receive a deferment for one year because he was a student. He explained his keen interest in parks and recreation studies and showed the board members copies of the *Sierra Club Bulletin* and other influential publications that highlighted major environmental issues across the country. The board members then excused him from the room and soon called him back. They had decided to consider his internship as a continuation of his education and give him a one-year deferment.

It was not an unmitigated success. Sansom drove to his parents' home in Lake Jackson—his mother was ecstatic, his father clearly not. The elder Mr. Sansom was a commander in the Naval Reserve, and having fought in World War II, he felt his son also should serve his country in the military. After the draft lottery was instituted, Sansom drew a high number and never was inducted. Still, the Vietnam episode had a lasting and profound effect on Sansom. For the rest of his life, he felt that he should devote himself to public service because of not going to Vietnam, and he held in high esteem those who did serve militarily. Moreover, he remained curious about the experience of soldiers who fought in a war that he opposed in principle.

In any case, the deferment cleared the way for his job in Washington, DC. And that move instantly elevated him into the national arena—in more ways than one.

Chapter 4

First Earth Day

In September 1962, biologist Rachel Carson published her seminal book, *Silent Spring*, which exposed how excessive use of pesticides was destroying the environment and endangering the health of people and animals. Groundbreaking in its time, the book showed that human activity can harm the natural world, and the bestseller eventually led to both a ban on the chemical pesticide DDT and the creation of the US Environmental Protection Agency. Equally important, it launched an environmental movement that was a major force in the social upheaval of the 1960s.

Some scientists in the '60s took the pessimistic view that Earth's long-term prospects were bleak. *Silent Spring* suggested that the planet's ecosystem was reaching the limits of sustainability. Paul R. Ehrlich published *The Population Bomb* and cofounded the organization Zero Population Growth to warn that the number of people on Earth could not increase if long-term environmental sustainability was to be achieved.

During the 1950s, environmental policy had not been a top priority in Washington, DC, though President Dwight Eisenhower had taken several small steps that paved the way for more significant ones later on. In 1958, he created the Outdoor Recreation Resources Review Commission. He followed that in 1960 with establishment of the Arctic National Wildlife Refuge. Several years later, the Land and Water Conservation Fund was established as a result of recommendations from the outdoor recreation commission.

By the early 1960s, environmental policy was climbing the political agenda. President Lyndon B. Johnson pushed legislation to protect the environment, having grown up in a rural area where he lived close to the land and water. He was born on a farm in the Texas Hill Country and later owned a ranch nearby. The first lady, Lady Bird Johnson, urged citizens to "plant a tree, a bush, or a shrub" to help keep America beautiful.

She became synonymous with her Beautification Project that focused on highways, parks, and neighborhoods. The president, for his part, signed landmark legislation to protect air, water, and land, and to create national wilderness areas. In 1964, Congress approved recommendations from the Outdoor Recreation Resources Review Commission to create a Land and Water Conservation Fund that both financed the acquisition of land and waterways for parks, and protected forest and wildlife areas. Suddenly, hundreds of millions of dollars were available for new recreation areas across the country. Overall, President Johnson signed more than three hundred conservation measures into law, forming the legal basis of the modern environmental movement.

In 1969, concern for the environment and alarm about pollution and ecological devastation surged. In January and February of that year, an oil spill in the Santa Barbara Channel, near the city in Southern California, sent three million gallons of oil into the Pacific Ocean. It was the largest oil spill in US waters at that time, and it still ranks third after the 2010 Deepwater Horizon and 1989 Exxon Valdez spills.

Then, in June of 1969, Cleveland's Cuyahoga River caught fire for at least the ninth time since 1868. The Cuyahoga had long been a dumping ground for sewage and industrial waste. But on June 22, 1969, a spark flared from the train tracks down to the river below, igniting an oil slick and industrial debris floating on the surface of the water. Not only did flames rage across the river, in some places reaching five stories high, but TV screens across the country captured the public's attention. Images of oil-coated birds, dead dolphins, and rivers on fire filled broadcast programs, newspapers, and magazines. Citizens responded by mobilizing cleanup campaigns and lobbying efforts against oil companies and other polluters. Demands for regulation of pollution could no longer be ignored. President Richard M. Nixon, suffering politically from criticism over the Vietnam War and a poor economy, formed a council on environmental quality, which led to the establishment of the Environmental Protection Agency in 1970. Ultimately, Nixon signed more environmental legislation than any president up to that point.

January 1969 was a perfect time for Sansom to arrive in the nation's capital, even if it meant without his wife. Nona stayed behind in Texas to finish her master's degree in education at Texas Tech, using a fellowship from the National Defense Education Act. Five months later, she packed all the couple's worldly goods into a Volkswagen Squareback, and Sansom flew to Lubbock so they could drive cross-country to Virginia together. A couple of days later they reached the Americana Fairfax Apartments in Annandale, Virginia, very late on a weekday evening.

"Exhausted, we went straight to bed," Nona recalled. "In the morning, Andy took the bus into the city to work, and I wanted to unload the car and begin to get settled in." Their apartment was on the second floor at the opposite end of the parking lot. "As I came down the stairs, I saw a precious little blond, curly headed girl playing in the grass in front of the apartment directly below ours. She looked to be between two and three years old. She jumped up and said to me, 'What's your name?' I replied and asked her name. She said, 'I'm Lizzy!'"

Thus began a lively conversation between the two of them as they unpacked goods from the car and carried them to the apartment. Little Lizzie wanted to help, so Nona put shoe boxes in her open arms each trip. "I think she carried our entire shoe collection into our apartment," Nona chuckled. "Lizzy told me all about her family—mom, dad, and new baby sister, Mary Claire. She asked me about my family, hoping, I think, that I had children who could be playmates with her. On our last trip, her mother called her into their apartment for lunch. We said goodbye and I went up to eat my lunch."

Shortly after lunch, Nona heard knocking on our door. "I opened it to see Lizzy and her mother, Lynne," Nona recollected. "I invited them in, and we spent a delightful time getting to know one another. We had both just finished degrees, she was doing final touches on her PhD in English Literature from the University of Wisconsin at Madison. She later asked me to proofread her dissertation for typos as her husband just didn't have time to do it. I was willing."

As it turned out, Lynne's husband and Lizzy's father was Dick Cheney, who later would be the power behind the throne of George W. Bush's presidency. In fact, he was the most powerful vice president in US history, according to many historians. At the beginning of his career, in 1969, Cheney worked as a staff member for the Office of Economic Opportunity under the directorship of Donald Rumsfeld. Dick and Lynne Cheney had grown up in Wyoming, and met and fallen in love in high school, as Lynne explained to Nona. "She was very congenial, interesting, and fun to be with," Nona said.

That summer, Lynne was at home taking care of an infant and a toddler as well as polishing up her dissertation. Nona was at home waiting for her teaching job to begin in September. "Thus, we had the chance to 'hang out' quite a bit," Nona recollected. "I remember going to the apartment swimming pool with Lynne and her girls. We would take our car to the park, or the mall, or other places to entertain Lizzy and MC. I also remember our families going on a picnic out in the nearby Shenandoah Mountains."

Thereafter, Sansom's path occasionally crossed with Cheney's. In 1980, both were delegates to the Republican National Convention where George H. W. Bush lost the nomination to Ronald Reagan. In the late 1990s, after Cheney had taken over the chairmanship of Halliburton, he was at a duck hunting club owned by the company near West Columbia and stumbled upon a book written by Sansom. The club was used for fundraisers by the Texas Parks and Wildlife Department, which Sansom was running at the time. "The manager of the club said, 'Well, Chairman Cheney was down here last week and he saw one of your books,'" Sansom recounted. "I sent him a copy and he sent me a real nice note about it."

Decades later, Liz Cheney became a "national hero" to the Sansoms. In 2021, the Republican congresswoman from Wyoming served as vice chair of the House Select Committee that investigated the January 6 insurrection and publicly contemplated a run for president. In October of that year, Cheney held a fundraiser in Dallas, and the Sansom's couldn't resist.

"The debate was not whether to go," Sansom explained. "But whether or not to contribute enough to be a VIP donor. We ultimately did that.

"I told her that our family had lived in the apartment above her family at Americana Fairfax," Nona recalled. "She smiled and said 'Yes, I've heard that story.' We think her parents must have seen our names on the guest list and explained to Liz who we were."

In 1969, Sansom's internship at the National Recreation and Park Association comprised two main responsibilities: organizing the association's regional meetings around the country and managing a jobs database to match students with suitable openings. Both played to Sansom's strengths. He knew how to plan and hold conferences, having learned logistics ranging from audiovisual equipment to marketing and publicity while at Texas Tech. And he excelled at human relations—he was intuitive, good at reading people and recognizing which relationships to cultivate.

Sansom began to push for a student counterpart to the main association, which soon became the National Student Parks and Recreation Society. Formation of the group enabled him to visit college campuses across the country and tap into the zeitgeist of the time. "There was a sense that change was in the air, of hope and change, that students could make a difference," Sansom recalled.

He sought out likeminded students in Washington, which wasn't hard given the capital's magnetism for idealists seeking change. Common Cause, a nonprofit dedicated to accountable government, had offices near Sansom's workplace and provided gathering space for environmental activists who were starting to plan a "teach-in" at college campuses. The planners had high-powered backing. Not least was Laurance Rockefeller, one of the richest men in America and a committed conservationist who funded the expansion of Grand Teton National Park and was instrumental in establishing and enlarging national parks in Wyoming, California, the Virgin Islands, Vermont, Maine, and Hawaii. Another wealthy Republican backer was Dan Lufkin, a cofounder of the investment bank Donaldson, Lufkin & Jenrette.

Then there was Democratic Senator Gaylord Nelson, an environmental champion from Wisconsin, who believed that antiwar sentiment on campuses could be tapped to demand action on ecological disasters. Nelson invited Pete McCloskey, a Teddy Roosevelt-type Republican who cared about nature, to join him in planning the event. "The objective was to get a nationwide demonstration of concern for the environment so large that it would shake the political establishment out of its lethargy," Nelson later said.

Nelson hired Denis Hayes, a Harvard student, to organize the "teach-in" event that was to be modeled after one held at the University of Michigan in March 1970. Hayes assembled a national staff of eighty-five to promote the event, joined forces with other organizations and faith groups, and changed the name to Earth Day. The intent was to engage audiences well beyond college campuses.

"Our theme was 'Give Earth a Chance,' which saw great publicity across the nation," recalled Arthur J. Hanson, one of the organizers, who was a doctoral candidate in ecology at the time. On the day, April 22, 1970, about twenty million Americans gathered, marched, and demonstrated to demand action against pollution and degradation of the earth. That was about 10 percent of the US population at the time.

During the preparations, Sansom got to know Hayes and worked closely with him. An environmental trailblazer, Hayes was the first of a lifelong series of coincidental encounters that Sansom had with historical figures who were shaping events of their time. These encounters proved serendipitous.

Hayes invited Sansom to speak at several Earth Day events, and Sansom addressed audiences at six campuses that day, the largest being the University of Illinois. At some of the stops, he was paired with New York State's first commissioner of parks and recreation, Sal Prezioso. Sansom showed graphic photos of the Santa Barbara spill and Cuyahoga River fire to his audiences. His call to action was, "Get involved, you have an obligation to protect the environment." The emerging activist crisscrossed the country by plane on that first Earth Day, bumping into Hayes in at least two airports.

Sansom's evolution fed off the ecological and political protests of the time, many of which were staged in Washington, DC, where he and his wife were living. "Nona and I would go down to the Mall almost every weekend and ten thousand or twelve thousand kids would be there," Sansom recalled. "It wasn't always the environment; it was also Vietnam. There was a wonderful atmosphere among the young people that was palpable. I've got pictures of Nona pregnant, sitting in a field of daffodils, in the midst of the demonstrations."

The times were heady for Sansom as he mingled with the likes of Hayes, Lufkin, and others. Yet they were tempered by life's reality. He had a wife and a baby due only two months after Earth Day. That meant a growing family who needed him—emotionally and financially. He couldn't carouse in bars all night planning Earth Day events and recruiting members for the National Society of Students in Parks and Recreation. Drugs, alcohol, and sex were readily available, but Sansom had other obligations. He needed not only to take care of his family but also to dress and behave in ways that would keep his professional options open for the future. He was mingling in circles with influential figures, such as Laurance Rockefeller and Lady Bird Johnson.

Sansom rode the wave but remained part of the establishment. His ability to push for change, sometimes radical—while firmly embedded in the establishment—became a lifelong trait.

Chapter 5

The Big Thicket

President Richard M. Nixon famously said that environmentalists wanted to "live like a bunch of damned animals" and were "interested in . . . destroying the system." Still, he knew a winning issue when he saw one, especially given that the Vietnam War and the economy were losing issues. Therefore, he pushed an aggressive environmental agenda regardless of what he might have thought about individual environmental activists. While the thirty-seventh US president is usually remembered for his Watergate scandal and ignominious resignation, his record on environmental legislation is a forgotten feat:

- National Environmental Policy Act of 1969
- Environmental Protection Agency in 1970
- Clean Air Act Extension of 1970
- Clean Water Act of 1972
- Endangered Species Act of 1973
- Safe Drinking Water Act of 1974

Ironically, the Clean Water Act—one of the most significant environmental laws in US history—is lumped in with those passed during Nixon's presidency even though he vetoed it. Only a congressional override saved the legislation.

In keeping with the tenor of the times, the White House in 1971 decided that its decadal Conference on Youth (originally the Conference on Children and Youth, held every ten years since President Theodore Roosevelt) should focus on the environment, among other things. Sansom, with a knack for being in the right place at the right time, was only too happy to help.

As an intern at the National Recreation and Park Association, he had gotten to know a *Christian Science Monitor* journalist by the name of Robert Cahn, who had won a Pulitzer Prize in 1969 for writing about the threats to national parks from overuse. Cahn was helping to organize the White House Conference on Youth through his membership on the newly created Council on Environmental Quality, and he asked Sansom to write a white paper for the environmental track of the conference. Sansom was thrilled to get involved in a national forum that addressed issues close to his heart and drew on the event-planning experience he had gained as a student and intern.

The administration wanted to capitalize on the second Earth Day by scheduling the event over the three days leading up to April 22, but it feared student protests if the conference was held in Washington, DC. Sansom suggested holding it in the western part of the country, far from large cities where demonstrators could more easily gather. Specifically, he recommended the YMCA of the Rockies in Estes Park, Colorado, set amid alpine beauty and located away from urban areas. While Sansom hadn't visited the resort himself, he had come across it while doing research he undertook as an ambitious intern.

It was a golden opportunity for Sansom. His white paper caught the eye of the new US Secretary of the Interior, Rogers Morton, a keynote speaker at the conference. Morton, a former congressman from Maryland, had been appointed by Nixon only three months earlier. He was somewhat of an anomaly as the first interior secretary from the East Coast. The position was typically reserved for prominent men from Western states. Morton had expected to be appointed earlier, in 1969, as a reward for working on Nixon's successful election campaign in 1968. But he was passed over in favor of a westerner, Alaska governor Walter Hickel. However, Hickel's tenure was cut short in November 1970 when Nixon fired him over a leaked letter that criticized the president for ignoring America's youth movement protesting the Vietnam War.

Not long after the conference, Sansom was invited to Morton's office. "I was actually terrified to go see a cabinet officer, but he asked me to come see him, so I went," Sansom recounted to Dr. Jen Brown of the Texas A&M-Corpus Christi Special Collections. "He and I had a pretty spirited conversation about my views and his on the environment." Morton then mentioned that he was due to give a commencement address the next

day at the University of Maryland in College Park and led Sansom to
another part of the office, where staffers were kneeling on the floor and
scrambling to paste together the speech. Morton asked Sansom whether
he could pull the address together for the University of Maryland commencement the following evening. Sansom had never written a speech
before, much less for a cabinet secretary. But that didn't stop him. He
called Nona and told her he wouldn't be home that night. He stayed up
all night and wrote the speech. Then he went home, took a shower, and
returned to the office. He handed Morton a draft early in the morning.
Morton marked it up, and Sansom typed a clean copy. Morton gave the
address on January 25, 1971, and Sansom started his new job as Morton's
speechwriter and personal assistant.

The speech was a tour de force. It encapsulated everything he believed
about environmental conservation and put it in Morton's words. The
speech set out most of the principles that guided the rest of Sansom's
environmental mission. It sounded more like a commencement message
that Sansom would like to have heard than a standard presentation to
graduating students. It was prescient.

The address started routinely with courtesies and then got to the heart
of the matter. "The great question of the seventies is, shall we surrender
to our surroundings, or shall we make our peace with nature and begin
to make reparations for the damage we have done to our air, to our land,
and to our water," Morton said, quoting President Nixon in his State of
the Union Address in January 1970. "Restoring nature to its natural state
is a cause beyond party and beyond factions. It is a cause of particular
concern to young Americans—because they more than we will reap the
grim consequences of our failure to act on programs which are needed
now if we are to prevent disaster later."

Morton went on to assure his audience that it wasn't just Republicans who were concerned about the environment—so were Democrats.
He quoted his predecessor, Stewart Udall, as saying, "As we enter our
third century, conservation must sharply define its concern for specific
resources and teach us to realize that plans to protect air and water,
wilderness and wildlife, are in fact plans to protect man. We will attempt
to enhance these resources not simply because it is desirable to have
wilderness or wildlife or clean air and water but because such action is
essential to the self-renewing systems of nature that sustain the earth."

By this time, Morton launched into a call for urgent action, asking who will "bring the preservation and management of our environment into a priority equal to that of our culture, our economy and our defense? There must be a change in our unwillingness to believe and understand the finite limits of the resources we extract from the earth."

Morton warned that the costs of conservation would be high. "The price of remedial changes in industrial processes, in the ecological impact of our works, in our use of land . . . and all the rest will be substantial." He explained that a change in values would be required. "There must be a change in attitudes of the citizen. They are necessary to make the whole proposition of conservation stick and become successful."

Exhortations to action were couched in pragmatism, an enduring trait of Sansom. "Our environmental equations must be achieved through the vigorous pursuit of the doable—that which we are able to do," Sansom wrote for Morton. "What I am saying is that what we have to do is doable—tough but doable."

Industry should be applauded when it does the right thing, Morton said, echoing a sentiment that Sansom held throughout his career. "We are witnessing a great change in the attitudes of industry," the interior secretary told the students. "We must give credit where credit is due." Sansom drew on another of his foundational beliefs—the importance of relationships. "It is not unreasonable for you to believe and understand that civilization is in fact an environment of relationships—man to his fellow man, man to his arts, and man to the rest of nature." The address ended with a look ahead to future generations, arguing that action was required out of respect for them. Sansom quoted Thomas Jefferson, saying, "not what was to perish with ourselves—but what would remain to be respected and preserved through other ages."

The speech was an inflection point for Sansom. It catapulted him to a national stage where he could give voice to his evolving views on environmental conservation through words spoken by the US interior secretary. Morton was a major player in Nixon's 1972 presidential campaign, having previously chaired the Republican National Committee. He traveled extensively, often taking Sansom with him. As they traversed the country, the energetic twenty-seven-year-old assistant wrote speeches on the fly, usually on a portable typewriter unless he was able to borrow an electric one at the Interior Department's regional offices.

Sansom's job extended to assisting Mrs. Morton as well. Anne Morton actively championed Interior Department causes, including the deterioration of infrastructure in national parks. She loved travel and cultural exchange, and Sansom became her traveling companion, taking care of logistics and setting up media appointments.

One of his early trips with Mrs. Morton was to Yosemite National Park, which at that time often drew antiwar protestors from San Francisco, a hippie haven only a few hours away. "There'd be thousands of kids on the meadow in Yosemite on any given weekend," Sansom told the Texas State University Oral History Project. "The typical park service culture at that time was to bang them over the head with batons. It was what we used to call heavy leather."

Given the growing power of youth culture generally and environment causes in California, a new park superintendent was appointed for Yosemite who took a different tack. "He recruited a group of young rangers who could relate to the kids," Sansom explained. One of the rangers was an acquaintance by the name of Walt Dabney, whom Sansom first had met when he was executive director of the National Student Recreation and Parks Society. When Sansom met Dabney the next time during the Morton visit, "he was playing a guitar out next to a campfire with some of the kids, and the whole thing had calmed down because they didn't beat on them, they related to them."

Another trip with Mrs. Morton was significant for Sansom as it resulted in the first conservation project for the budding environmental conservationist. Until the 1970s, wolves were largely considered dangerous and rapacious killers in the United States, reflecting the demonic symbolism and hostile view of predators that had pervaded Western culture for centuries. Anne Morton, however, was a devout advocate for their preservation. And who was better suited to help her than Rogers Morton's special projects man, an eager and sociable one at that? The duo traveled north to Alaska, where they met with biologists in Mount McKinley National Park (now Denali National Park and Preserve) researching two packs of wolves that had been under study since the 1940s. The packs lived close to two rivers, the Savage and the Teklanika, ranging throughout the river basins, which were partly inside the park and partly outside. The wolves were protected on public land until they crossed the line. Then they could be killed at will, including shot from helicopters under Alaskan regulations at the time.

Sansom chartered a Piper Cub airplane and followed the wolves for two weeks, photographing them to prove they were moving in and out of the park, providing previously lacking empirical evidence for the Interior Department. Consequently, wolf habitat areas were added to the park to protect the animals.

There was an epilogue, too. The Mount McKinley park rangers decided to take Mrs. Morton on a dogsled trip. After packing her up in caribou skins, with her face barely visible, a ranger stepped onto the back of the sled and off they went, plowing through five feet of deep snow. Sansom's job was to follow and photograph her adventure. Soon the sled stopped so that Sansom could get a good shot. When the ranger stepped off the sled to ask Sansom if he was finished, Sansom said, "Okay."

But that was exactly the command the dogs had been taught as their signal to start running. And run they did, with no driver and no control. "Both of us looked at each other and said, 'Our careers are finished,'" Sansom recalled. The last thing Sansom saw was the sled disappearing over a hill with the ranger trying to run through waist high snow to rescue Anne Morton.

All was not lost. Morton kept her wits about her. When the sled rounded a curve and sped downhill, she escaped by rolling out of it. In the meantime, the ranger ran back to headquarters by a different route, thinking the sled, the dogs, and Morton would arrive soon. The sled and dogs did arrive, though not Morton. The ranger figured she was back along the trail somewhere. He leapt back onto the sled and said, "Okay." The dogs bolted although they still were tied up, sending the ranger flying over the sled and into the dogs. A few minutes later, Morton came trudging back, and the slapstick comedy made its way into the *Washington Post*.

There was a postscript to the epilogue. Interior Secretary Rogers Morton also was visiting various sites in the state, using the department's Learjet to get around. His itinerary included Prudhoe Bay, where vast oil fields had been discovered three years earlier. The secretary, however, got called back to Washington, DC, for an emergency cabinet meeting and took the jet with him, leaving Sansom and Anne Morton high and dry.

"Watergate was just breaking about this time, so chaos reigned," Sansom recalled. He called Western Airlines and discovered that if you had a round-trip ticket from San Francisco to Anchorage, you could also fly to Honolulu for free.

"So I went and knocked on Anne's hotel room door. Her daughter was with us at the time," Sansom recalled. "I said, 'Ladies, we're going to Hawaii.'" They loved the idea. Parks were a passion of Anne Morton, who served on the National Park Service Advisory Board from 1975 to 1981 and on the World Wildlife Fund Board from 1983 to 1986. The trio jumped on a plane the next day and hopped over to Hawaii, where they visited Volcanoes National Park.

"They were all wondering where the hell we were back in Washington," Sansom recalled jokingly, though in actuality, he had advised the secretary's office of their change in travel plans.

Another foray with Anne Morton to Alaska also tested Sansom's resourcefulness. It involved a trip to Mount McKinley. The hotel where they were to stay had burned down on Labor Day of 1972, and the concessioner planned to have it rebuilt by the following Memorial Day in time for the summer high season. But construction in six or eight feet of snow was tough. Moreover, rebuilding the hotel was opposed by some environmental groups, which argued that it was inappropriate in a national park. The delays deep-sixed the construction schedule.

Undaunted, Sansom and Morton hatched a plan. Together with the Mount McKinley staff and concessionaire, they devised a way to build a hotel in only nine months that would accommodate park guests. The concessionaire bought train cars from Alaska Railroad, which ran through the park, and strung them together on rail sidings that already existed. "If you got a room in that hotel, it was in a sleeping car parked on the siding," Sansom explained. There was a dining car, and the only structure that had to be built was the registration desk, which looked like a train station. The Mount McKinley Station Hotel opened in 1973 as planned, continuing in operation for twenty-eight years despite originally being considered a temporary solution.

Sansom's adventurous travel and late nights at the office contrasted sharply with the domestic life that was Nona's world at the time. "Ask any young mother, it's the most stressful time," Nona recalled. No family members or in-laws were available to babysit, as they were back in Texas.

"Andy was working as a special assistant to the secretary, and so he did a lot of traveling. He was often at the office late at night, too. I had a parttime job just to keep my sanity, but I had to make sure that dinner was ready in the evening."

The forays to Alaska were productive and good fun and launched Sansom's lifelong love affair with Alaska. By his reckoning, he has been to Alaska more often than any other state except those where he and his children live.

Still, Sansom was officially Morton's speechwriter, and churning out presentations for someone else to deliver wasn't Sansom's forte. In his own words, he "put too much of his own voice" in Morton's speeches. It was a tenuous time, and Sansom was only too happy to get a surprise call from the State Department. US embassies around the world, especially in Asia, wanted a youth envoy to come and talk about the American environmental movement to various audiences in their countries.

Sansom asked his boss whether he could accept the State Department's invitation to tour Asia and Australia for three months. Morton approved wholeheartedly, though there was still the question of whether Nona, with a toddler under the age of three to care for, would agree. She did, though she and little Andrew took refuge with her parents and in-laws back on the Texas coast while Sansom was away.

From January to March of 1972, he traveled much of Asia between South Korea and Australia. One of his most memorable speaking stops was Singapore, the city-state on the Malay Peninsula. It was run by Prime Minister Lee Kuan Yew, who was seen by some in the West as authoritarian for curbing civil liberties in order to promote economic growth. In the late 1960s, his administration started discouraging long hair on males because it was associated with Western decadence, specifically hippie culture. Male travelers who arrived with long locks—reaching below the collar, covering the ears and forehead, or touching the eyelashes—were barred from entrance or had their passports confiscated until they got a haircut.

Sansom's hair at the time reached his shoulders. He only got in because he was carrying an official US government passport—and he was not forced to get a haircut. However, he was not allowed to do a TV interview that had been scheduled because of his appearance. Instead, a radio interview was substituted. Then the itinerary got more interesting. US

Embassy staff members explained to Sansom that he would be addressing the Central Council of Malay Cultural Organisations the next day.
"Malays are discriminated against here like Blacks are in America," he
was told, with Chinese being the predominant ethnic group. When Sansom asked why they wanted to hear from him, he was told, "The council
will not allow people from the embassy to address them." Minority
Malays were skeptical of the United States after it recognized Singapore
in 1965 following expulsion of the city-state from the Malaysian Federation. In 1969, the Central Council of Malay Cultural Organisations was
created to promote the interests of the predominantly Muslim Malays.

The US Embassy staff arranged to drop Sansom off on the street to
meet the event hosts, who were in charge of returning him to the hotel
afterward. The audience of Malay men, puffing on cigarettes in a smoke-
filled room, listened politely while Sansom made his pitch about the
environmental movement in the United States. "And then they said,
'Okay, we want to hear about Martin Luther King.'" What they really
wanted to know was how the civil rights movement in the United States
was working because that was what they were facing in Singapore.

Eventually, Sansom arrived in Australia, and then he went rogue. He
had a few days until his next scheduled stop in Japan and decided to
detour to Vietnam, which was not on his itinerary. The war represented
an emotional conflict for Sansom, who was relieved not to have been
drafted but vaguely ashamed that he had let his father down. He flew
Pan Am to Saigon, the only civilian on the plane crowded with military
personnel returning from R&R in Manila. "It was the most morose flight
that I can ever remember being on," Sansom recollected.

When he got to Saigon, the passport officials did not know what to do
with him because very few American civilians were arriving. He had a
red passport, though, which was the official US government passport,
and they let him through. He stayed in Saigon for about three days, amid
audible shelling and bombing.

On the first day, he walked down the street looking for a place to
change money and spotted a GI crouched in the corner of an Asian ar-
cade smoking a cigarette. Sansom asked him where to go, and the soldier
offered to take him to the USO, which Sansom happily accepted. He got
the Vietnamese money and thanked the soldier, offering to treat him to
dinner. Sansom wanted to put most of the money in his hotel room safe,
so he invited the solder to go with him to do that.

When they got to the room, the soldier took his cap off and his hair fell down below his shoulders. When Sansom asked him how he could get away with that, he said he had not been to his unit in four months. Over the next few days, the soldier introduced Sansom to a group of other GIs who wore uniforms but stayed underground.

After three months, the Asian odyssey ended and Sansom arrived back in Washington, DC, on a Saturday. "After this magnificent trip, I got reacquainted with Nona and Andrew," his young son. Sansom admitted that "Nona was the anchor of our household and carried most of the load in those years."

By the next morning, however, he raced to the Interior Department, hoping to find various things he had sent back to his office by diplomatic pouch. Arriving at the corner office overlooking Constitution Hall, Sansom was shocked to see his nameplate replaced by another one.

He looked around anxiously and finally found his name on a window-less office by the men's restroom. "In the DC environment, they learn to do without you pretty quickly," Sansom said. Although he still had an office, it dawned on him that he didn't really have a job anymore. He knew he wasn't a great fit as a speechwriter, and he needed to support his young and growing family, whom he hadn't seen in three months. He had to scramble.

Once again, Sansom's instinct for cultivating valuable relationships came to his rescue. He had befriended an assistant secretary at Interior named Nathanial "Nat" Reed, who was assistant secretary for Fish, Wildlife and National Parks. When Reed was introduced to the president in 1970, Nixon told him he didn't give a damn about environmental issues. He had too many things on his plate. Nevertheless, Nixon asked Reed what his priorities were, and Reed responded that he wanted to ban the pesticide DDT and the chemical known as Compound 1080, a poison that killed coyotes and other predators, as well as many untargeted animals in the West. As a result of Reed's rapport with Nixon, possibly due to their mutual love of Florida, when the Environmental Protection Agency was created that year, it banned all uses of both.

Sansom attributes much of Nixon's environmental accomplishments to the influence of Rogers Morton and Nat Reed. When Sansom was dismissed from Morton's office as a speechwriter in 1972, he ended up in Reed's shop, still at the Interior Department and still at the beck and call of Morton. He and Morton remained friends for decades.

Sansom's biggest assignments under Reed were in his home state of Texas, and unlikely patches at that. The first major project involved some 150,000 acres of swampy land in East Texas known as the Big Thicket, the state's wettest, most densely forested ecosystem. It provides a home for 500 vertebrates, more than 160 species of trees and shrubs, 800 herbs and vines, and 340 types of grasses. Unlike most places chosen for national parks, the Big Thicket offers no majestic peaks or sweeping views. Its attraction is the immense diversity of soils, plants, and animals and the relationship between them. Carnivorous plants feed on animals, and eleven species of bats prey on more than 1,600 species of moths. As many as eleven ecosystems exist amid the boggy bayous, hardwood bottomlands, and pine forests, earning the Big Thicket a UNESCO designation as a biosphere reserve in 1981. But long before it earned the recognition it deserved, a group of dedicated and tenacious activists began the protracted battle to preserve a place that even other Texans thought better belonged to the alligators, snakes, and mosquitoes.

One ardent advocate was Ned Fritz, an attorney and environmentalist from Dallas. Author of the book *Realms of Beauty: The Wilderness Areas of East Texas*—a book that would introduce the Thicket to new audiences—Fritz traipsed the halls of the capitol in Austin and trekked to Washington, DC, to lobby for making the area a national park. He had flaming red hair and talked a mile a minute, according to Sansom. When he was excited, he often began every sentence with, "Lookie."

Another evangelist was Geraldine Watson, a self-trained botanist who could identify and describe virtually every plant in the Big Thicket. At one point in their long, frustrating fight to preserve the thicket, Watson and others in the fight thought about giving up on the project, saying their time could better be spent elsewhere because the politicians were never going to act. But Fritz wouldn't let them off the hook. Sansom remembered that Fritz would call Watson any hour of the day or night and say, "You can't give up on this. You got to keep working."

The biggest challenge for those who wished to preserve the Big Thicket was its history of exploitation. Around 1901, oil wildcatters descended following the birth of Texas' oil industry at the Spindletop gusher in deep Southeast Texas. The oil boom in the wooded wetlands lasted only until about 1908, though it gave way to a much older and bigger industry—timber. Being the most densely forested part of Texas, logging began there in the early 1800s. The arrival of the railroad helped fuel the industry, and by 1907, Texas ranked third in lumber production among the states, most of it coming from the Big Thicket.

When efforts to preserve the area started in the 1920s, it covered as much as two million acres. By the 1970s, when Sansom entered the picture, logging had pared that back to a couple of hundred thousand acres. Most timber companies opposed preservation of the swampy forest. One of the companies that was more supportive was Temple Industries, which traces its East Texas roots back to 1893 and today is part of International Paper Inc.

Temple was less opposed to protection because its property would have been least affected, some said, being farther north than other major companies. But Sansom attributed much of Temple's position to the enlightenment of Arthur Temple, who played a major role in preserving the Big Thicket. Another major player was Pat Noonan, president of The Nature Conservancy, who—along with Sansom—had been mentored by Robert Cahn of the *Christian Science Monitor.*

By the early 1970s, Temple (and other timber companies) had quit logging in the area of the proposed park. Temple had stopped clearcutting, razing cutover lands, and using airborne chemicals to kill underbrush. In addition, Temple was moving toward slower-growing trees such as oaks and away from faster growing ones such as pines, which were favored by other companies.

Not unimportantly, Temple also had the corporate backing of publishing giant Time Inc., which became its parent company in 1973. Time acquired Temple and merged it with its Eastex Pulp and Paper Company, which had owned timberland in East Texas since the early 1950s. Arthur Temple persuaded Time's top management back in Manhattan to support the park project.

Sansom already knew about the fight for preserving the Big Thicket from one of his environmental heroes, Senator Ralph Yarborough. A New Deal Democrat, Yarborough grew up in the Piney Woods of East Texas, which encompass the Big Thicket and are part of a larger coniferous forest stretching into Louisiana and Arkansas. Yarborough had championed a national park in the Big Thicket since the mid-1960s. The Democrat was an old-school, liberal politician who ran on the slogan, "Let's put the jam on the lower shelf so the little people can reach it." In 1966, he submitted a bill to set aside seventy-five thousand acres to create the Big Thicket National Park. The bill languished until Yarborough was defeated in the primary by the more conservative Lloyd Bentsen. Yarborough's last year in Washington coincided with Sansom's arrival in 1969, and they became allies.

A year later, in 1970, the Nixon administration had pledged support for the national park as part of its endorsement of George H. W. Bush, who was running for the US Senate after serving as a congressman. During his congressional tenure, Bush had introduced legislation to preserve the swampy treasure, as did several other Texas congressmen who had also introduced their own bills. Once the Nixon administration entered the fray, though, it fell to Interior Secretary Morton to get it done. Morton, in turn, made Sansom the administration's point person on the Big Thicket as one of the special projects occasionally assigned by the secretary. His task was to follow the issue on a day-to-day basis. And there was a lot to follow.

For one thing, the Charlie Wilson, the iconic congressman from Lufkin, Texas, deep in the Piney Woods, was involved. Wilson was a colorful political player, as depicted by the actor Tom Hanks in the movie *Charlie Wilson's War.* Wilson had worked for Temple Industries before going into politics, leading to his nickname "Timber Charlie." The more memorable nickname, "Good Time Charlie," was chronicled in news reports, Justice Department investigations, and the popular film. By his own account, Wilson was a "ladies' man" who liked buxom "Angels" as staff members. His drinking bouts and alleged drug use were part of the mystique that enabled him to take his job seriously without taking himself seriously, as some pundits said.

Sansom worked closely with Wilson on the legislation for several years, with the support of Buddy Temple, who measurably improved his company's environmental record along with parent company Time Inc. Eventually, their work and that of all the Big Thicket advocates bore fruit. In 1974, the piney baygall was designated a national preserve named the Big Thicket National Preserve, covering 84,550 acres. It was one of the first two preserves established by the National Park Service, the other one being the Big Cypress National Preserve in Florida. National preserves are similar to national parks except preserves allow extractive activities such as hunting, fishing, and mineral production if they don't jeopardize the "natural values," according to the National Park Service.

By 1974, Sansom had moved on, but the legislation was a departure for the National Park Service. The Big Thicket didn't fit the park service's traditional definition of a national park. The land had been logged, bottomlands broken up into fragmented pieces, and the water was swampy—not big, white, and rushing. Critics saw ragged clumps of trees, while advocates saw a "string of pearls." In the end, the Interior Department led the charge to create the preserve, not the park service.

For Sansom, the Big Thicket was a personal epiphany. He started realizing the difference between a park meant for recreation and a preserve meant for protection. Parks benefitted people, while preserves benefited nature. Recreation inserted humans into the homes of other species, while preservation protected those homes. The distinction was crucial and became the source of decades of handwringing and conflict within the conservation movement.

Public funds would be needed to acquire and maintain natural habitats for the primary purpose of ensuring their survival rather than providing entertainment for visitors. Greater political support was required if environmental policy was going to evolve, Sansom recognized.

"The whole idea of spending federal dollars and using federal management resources simply to protect biodiversity was in its infancy," Sansom recalled. It was a revelation that would shape Sansom's view of the world for the rest of his life.

Well before the bill was passed, however, Sansom got pulled over to another big, thorny Texas project—the second assignment in his home state.

Chapter 6

Matagorda Island

Like many countries, the Republic of Texas in the late 1830s wanted to protect itself against border incursions—mostly from Mexico. A string of barrier islands along the Texas coast proved useful in that regard. By 1843, the republic had succeeded in building a small fort on Matagorda Island—one of seven islands in the string—to defend its southern border.

For the next one hundred thirty years, Matagorda Island provided a conveniently remote spot for various military endeavors by Texas, the Confederacy, and the United States. No public bridges or causeways connect the mainland to the island, ensuring isolation. The absence of power and water utilities also warded off unwanted visitors.

The hot and humid island did attract an array of wildlife though, including waterfowl, alligators, and coyotes. The thirty-eight-mile stretch of sand dunes and salt grass is home to nearly twenty state or federally listed threatened or endangered species, including whooping cranes, the tallest and among the rarest birds in North America. The migratory birds spend their winters on Matagorda Island and the mainland across from it—known as the Aransas National Wildlife Refuge—after flying more than 2,500 miles from their breeding areas in Alberta, Canada. The island is a resting spot for many other birds that migrate between their nesting and wintering areas along the North American Central Flyway.

Matagorda Island's role in bird migration was of keen interest to the National Audubon Society as early as 1898 when a Texas chapter was created to protect waterbirds. In 1971, the society became alarmed about new offshore oil leases that the state was offering near Matagorda Island. While the state routinely offered fresh leases each year, those in 1971 took on particular sensitivity in the wake of the 1969 oil blowout of an offshore oil rig in Santa Barbara Channel. The birders weren't the only ones upset about potential oil rigs near the island. The US Air Force was equally unhappy. It had been operating on the island since 1942, including clandestine air-to-ground bombing training during the Vietnam War, and didn't want inquisitive eyes on nearby oil rigs.

The uproar over the offshore leases came to the attention of the Interior Department, which was becoming more attuned to environmental issues amid the Nixon administration's passage of landmark legislation. Secretary Morton summoned Sansom, who remained on-call for special projects while working for Reed. Sansom had gotten something of a reputation for mediating tensions between the US military and various Puerto Rican islands where it was operating. Not long before the Matagorda flareup, Sansom had been involved in negotiations between the Navy and villagers on the island of Culebra, which belonged to Puerto Rico. The Navy had seized most of the 134-square-mile island to use as an outpost for its Roosevelt Roads Naval Station on the Puerto Rican mainland. Culebra provided target training for air-to-ground bombing using live ordnance, among other things.

The military had allowed people to return to the island, but jets continued to fly overhead at four hundred or five hundred feet, disrupting school during the day. By the time Sansom came along, these flight missions had become controversial, and he helped facilitate an agreement among the parties to improve the situation.

The furor over oil leases near Matagorda Island gave Sansom, in his midtwenties and eager to please bosses, an opportunity to demonstrate his knowledge of the area and skill in finding common ground between quarreling parties. Sansom learned about the island's environmental significance from a report done by Texas State Senator Don Kennard, who was an early advocate of conservation in a state where protection of natural resources often was trumped by private property rights and laissez-faire policies. One of Kennard's lasting legacies was a natural area survey conducted at his behest by the University of Texas to document biologically and culturally rich areas of the state that might eventually be protected by the government. Matagorda Island was one of them.

When Secretary Morton asked Sansom what he knew about the island, Sansom regurgitated Kennard's report. Morton was convinced that Sansom knew what he was talking about and assigned him to fly down, do on-the-ground research, and make recommendations regarding the oil leases.

Sansom found an ornithologist in the Interior Department who was a leading expert on whooping cranes to accompany him on a field trip. He contacted the Air Force, which still operated the Matagorda Island Air Force Range. Officials there tried to dissuade him from the mission, but finally it was agreed that an Air Force minder would accompany Sansom and his whooper expert. The trio headed down to Matagorda.

The day they arrived, another plane also landed on the island, an aircraft with American military generals from Vietnam who had been flown in to hunt deer, quail, and ducks. Sansom's entourage mingled with the generals, who were staying at the island's hunting lodge.

Sansom and his colleagues spent three days on the island and determined, first, that increased offshore oil and gas activity would not endanger the whoopers. In addition, they concluded that the island was primarily a playground for military brass. What was less clear was the impact of the military operations on the birds. Training exercises included B-52s flying in from across the country and firing radar beams at the ground, though not live bombs. The bombing practice posed little risk to the cranes, though Air Force helicopter training and recreational hunting did, according to Sansom's report, which was debated in Congress. "They didn't need a remote island in the Gulf of Mexico to do that," recalled Sansom.

Sansom argued that military activities should end because they were stopping expansion of the whooping crane population, and he recommended the island be made a national wildlife refuge. He also organized allies, including the media, to support his position. Sansom had started learning about the power of the press after his arrival in Washington, DC. He was mingling with the likes of Pulitzer Prize winner Robert Cahn, who arranged for Sansom to write a conference paper that got him a job working for a cabinet secretary. Cultivating relationships with reporters and editors was a habit that Sansom nurtured for the rest of his life.

The owner of the Corpus Christi newspaper, the *Caller-Times*, was Ed Harte, scion of the Harte-Hanks Newspapers dynasty. Harte was captivated by the Matagorda Island controversy personally and professionally. The story was in his newspaper's bailiwick, and it involved whooping cranes, of special significance to him as an active member of the National Audubon Society. Harte's decision to consistently cover the whooping crane story in the *Caller-Times* was a bold and forward-looking move. The paper's position was described by the *New York Times* as "an unusual stance for a Texas newspaper at the time."

Sansom and Harte became acquainted and devised a plan in which Sansom would leak the Matagorda Island report to the Audubon Society if the Department of Defense didn't respond appropriately after receiving it from the Interior Department. Sansom hand delivered a letter from Secretary Morton to Defense Secretary James Schlesinger, who coincidentally was a birdwatcher. It was the Wednesday before Thanksgiving, and no answer came for several weeks after.

By the Christmas holidays of 1972, Sansom and Harte decided the right response was not forthcoming, and Sansom gave his report to Audubon as they had planned. The next thing he knew, he got a call from *60 Minutes*, the preeminent weekly TV news program of the time, asking for an interview. "Oh shit, I'm in too deep at this point," Sansom thought. He declined to be interviewed, but the story ran on *60 Minutes* anyway.

It was a big story, and the media ran with it. A few days later, Sansom was commuting to work in DC from Virginia on the bus. "I never will forget opening the *Washington Post* on the bus and seeing a half-page photograph of the whoopers on the front page," Sansom said. "I remember, on the one hand, feeling exhilarated. But, on the other hand, thinking I'm doomed."

Sansom knew his days were numbered at the Interior Department, and he harbored thoughts of returning to Texas and running for elected office, specifically Congress. Politics wouldn't necessarily offer stability, though the idea remained enticing. It was enticing enough that Sansom eventually asked Morton to put in a word for him with George H. W. Bush, then the head of the Republican National Committee. Morton and Bush were close, having served in Congress together and having chaired the Republican National Committee within a few years of each other. Morton was as good as his word, and a meeting was arranged between Sansom and the fifty-one-year-old Bush, also a former United Nations ambassador and director of the Central Intelligence Agency.

It was August of 1974 and President Richard Nixon was in the final throes of the Watergate scandal. "I go to the RNC headquarters, and it had been thrown it into such disarray by Watergate that it was chaotic,"

Sansom recalled. "I remember sitting outside Bush's office with people running back and forth like crazy, in the midst of this crisis. He finally came out and said, 'I'm so sorry. You know, this is a very, very difficult situation.' He asked me where I was going." Sansom answered that he was going back to the Interior Department, and Bush offered to give him a ride, driving an "old purple Gremlin." He wasn't being ferried around in a limousine, and that impressed Sansom. "He drove me back and interviewed me and encouraged me to come home and get involved in politics."

Nixon resigned within weeks.

Not long after that, Sansom was summoned to a meeting with one of the interior secretary's assistants named Roy Hughes, who was known as the "heavy" in the department. "'You're done,' he told me. 'This really isn't the place for you,'" Sansom said. Some years later, Hughes—who had gone to work for Enron by then—told Sansom that the Defense Department had exerted heavy pressure on Interior, and perhaps not surprisingly, Sansom took the fall.

Still, Sansom and Morton continued to play paddleball often. They remained friends until Morton's death, and Sansom attended Ann Morton's ninetieth birthday party at the Bellagio Hotel penthouse in Las Vegas in 2007.

Sansom lost his job, but he could claim an environmental victory on his resume. The previous year, in 1973, the endangered whooping cranes had gained a bigger home on Matagorda Island as the Air Force began transferring its nineteen thousand acres to the US Fish and Wildlife Service. Even though military operations continued for another decade or so, the exit paved the way for even more protection later when Sansom was in a position to wield power as the head of Texas Parks and Wildlife Department.

Sansom's sacrifice hadn't gone unnoticed. Ed Harte told him that "you've fallen on your sword for these whooping cranes, and the Audubon Society will take care of you." Harte followed through when he became chairman of the national board in 1974 after having been a member for a decade. Harte's offer of an Audubon job would have meant a move to New York City and a detour from Sansom's dream of running for public office in Texas, so he passed. Still, the gesture solidified the friendship between the two men.

While Sansom passed on a permanent move to New York City, he did accept a temporary assignment in Puerto Rico in the autumn of 1974 after leaving Interior. Earlier, while still at the department, he had visited Mona Island—located to the west of Puerto Rico, midway to the Dominican Republic. The hot, dry climate and geographical isolation nurtured species found nowhere else, such as the Mona Iguana, a reptile that later was designated an endangered species.

In the early 1970s, Puerto Rico had started thinking about building a deepwater port on Mona to transfer oil from supertankers coming from the Middle East to smaller vessels that would continue to the US mainland. "I was concerned about it while I was in the Interior Department," Sansom explained. "In fact, I was very vocally opposed to it. And I think that was another thing that ultimately got me in trouble at Interior."

In any case, Sansom was hired as a consultant by the Puerto Rican government to study Mona and make recommendations on how the island should be managed. The consulting gig was only a couple of months, yet it illustrated one of Sansom's most enduring traits. He was adept at finding ways to get paid to pursue pet projects that caught his fancy. Even when the projects were scrapped, the consequences for his career were rarely dire. And sometimes they were a boon.

Chapter 7

Don't Be Fuelish

In 1974, Sansom had no job and a wife plus two toddlers to support. His research project on Mona Island for the Puerto Rican government had ended with nothing else lined up. Nona was working part-time as a recreation director for a senior citizens center about fifteen miles from their home in Falls Church, Virginia. Their son, Andrew Jr., was a four-year-old who had been born shortly after the first Earth Day in 1970. Their daughter, April, was one year old.

What Sansom did have was a good friend who was well connected in DC circles. Dick Cheney had been an ambitious and energetic staff member at the Office of Economic Opportunity in 1969, when he and Sansom met. Cheney and his wife, Lynne, were the first friends the Sansoms made when they moved to the DC area. They socialized and babysat each other's children, including the Cheney's daughters—Lizzy and Mary.

In the days when Sansom still had a job, he and Cheney sometimes carpooled into DC in Cheney's Volkswagen Beetle, which had a broken passenger window. If it was rolled down, it got stuck in that position, and Cheney made sure no one did that. "He had taken pantyhose, filled them with briars, and wrapped them around the window crank," Sansom recalled. "So if you inadvertently grabbed the crank, you got a fistful of thorns and you let go right quick."

By 1974, Dick was deputy assistant to President Gerald Ford, and the Cheneys had left the rental apartment for a larger townhouse. "We saw the Cheneys only occasionally," Nona recalled. "In the mid-'70s, they invited us to join a group of their friends to sit in the President's Box at the Kennedy Center for a ballet performance. By that time Dick was chief of staff for Gerald Ford and the Cheneys had a whole new circle of friends."

Cheney maintained close ties to the Cost of Living Council, where he worked before moving into the administration and offered to put in a word for Sansom. The council was created to establish goals for the Economic Stabilization Act of 1970, and Sansom wasn't sure that controlling wages and prices to fight inflation was a good fit for him. He had met some of the staffers and noted that a few of them moved to the Federal Energy Administration when it was created in June 1974 to address a growing energy crisis more directly. That agency seemed a better fit for Sansom.

The new agency was looking for someone knowledgeable about energy conservation and communications. Sansom needed a job, and he could claim expertise in "conservation." Still, he was worried about one thing. He might be viewed as a troublemaker, given his divisive report on Matagorda Island.

"I did not want that to be a surprise," Sansom said, knowing that Congress was still considering his recommendation to close the Air Force base on Matagorda and make the island a wildlife refuge. In a job interview with the assistant administrator at the energy administration, Sansom told him about the Matagorda report and "he loved it."

Conserving energy was a top priority for the agency in the mid-1970s. One of Sansom's projects involved taking an educational roadshow to selected audiences across the country. Drawing on his previous experience in organizing events, he partnered with the Conservation Foundation (today part of the World Wildlife Fund), and they put together a series of seminars for conservation groups, elected officials, and other interested parties. The purpose was to explain conservation-minded building technologies and to motivate audiences to be early adopters of it. "We were doing this by the seat of our pants," he admitted.

During one of the presentations in Texas, Sansom met Pliny Fisk III, a pioneer in green building technology who used environmentally conscious design and materials for his methods and projects. Fisk was both an architect and landscape architect, based in Austin. In 1975, he founded the Center for Maximum Potential Building Systems, a nonprofit devoted to research and design. His innovative ideas about sustainability and construction made a lasting impression on Sansom.

In his conservation and communications role for the Energy Administration, Sansom was drawn to the advertising side of the job. He was put in charge of the "Don't Be Fuelish" campaign to promote energy conservation that already was underway. His job was to expand the campaign from print to TV advertisements, working directly with the ad agency on the creative content. At one point, he suggested that Peter Max, an iconic pop artist of the 1960s and '70s, do the artwork for some ads aimed at the youth market. Max's psychedelic style using bright colors had made him a cultural phenom. He had created the first Earth Day poster in 1970 and the first stamp with the designation "Preserve the Environment" for the US Postal Service in 1974. Sansom was eager to meet him in person and arranged to visit Max at his Manhattan apartment. "He was all over it," Sansom said of Max.

Sansom's role on the "Don't Be Fuelish" conservation campaign led him all the way to the White House before politics engulfed one controversial ad. As a young communications staffer at a new agency, Sansom was assigned to take the ad concept to President Gerald Ford for review. It showed a chess board with Statues of Liberty as white pieces and oil derricks as black pieces, which were being moved by a figure in Arab robe and headdress. "I took that storyboard and showed it to President Ford in the cabinet room and he loved it," recalled Sansom. While the president approved, however, others hesitated. The ad was scrapped over concerns that it portrayed Arabs manipulating the United States.

By 1976, Sansom realized he needed to make a decision about his career because the Ford Administration was coming to a close. As a political appointee, he needed to either become a permanent government employee or move on. He decided to apply for permanent service even though he was facing stiff competition from returning Vietnam vets, who were given preference for civil servant positions.

In true Sansom fashion, he lined up an ally to shepherd along his effort—another next-door neighbor in Annandale. The neighbor worked in human resources at a federal agency and was able to monitor the application process. The paperwork reached the final hurdle, a signoff by Sansom's boss who had hired him at the Energy Administration, when a major roadblock appeared. The boss who was friendly to Sansom had left and a new one had taken over—one not disposed to approve the application.

The effort failed and Sansom weighed his options, feeling the pressure of supporting his young family while facing possible joblessness—yet again. He contemplated a return to his Texas roots, which had been the plan from the beginning. His children were now three and six years old, and he wanted more certainty that he could financially care for them. Being closer to family, his parents and Nona's, could provide support—and did, in ways that he fully understood only later. In terms of his career, Sansom was seeking more continuity.

In any case, Sansom did what he always did when he needed help. He mentally scanned his database of professional connections and remembered a University of Houston professor who was an expert in thermal radiation, or the transfer of heat between objects that are not in contact with each other. Radiative transfer was drawing attention in the field of energy-efficient building, such as "smart windows," and Jack Howell had written the textbook *Thermal Radiation Heat Transfer*. Sansom had met Howell during one of the educational seminars staged by the Federal Energy Administration.

At the time, Howell had wanted Sansom to come to the Energy Institute at the University of Houston, which focused on energy conservation and alternative energy. Now Sansom was more open to it. The institute promoted energy *efficiency* in a city that promoted energy *consumption*, being the "energy capital of the world." But that didn't stop Sansom. He figured he might be able to continue his work in energy conservation if he found the right fit.

Howell wasn't the only acquaintance urging him to return to Texas. Another was a grassroots activist who was a Lone Star State version of Erin Brockovich, the California environmental advocate who accused Pacific Gas and Electric Company of contaminating groundwater and won a $333 million settlement. The Texas version was Sharron Stewart, and coincidentally she lived in Lake Jackson, where Sansom grew up. She was a self-taught environmentalist who had battled the surrounding chemical companies over air pollution following a mysterious illness that she and a daughter had suffered. Stewart believed the sickness was related to air emissions from the chemical plants. From there, she took on water pollution in the Brazos River, where Sansom had spent much of his childhood. "The lower reaches [of the river] from Lake Jackson through Freeport and to the Gulf of Mexico were bottle green," Stewart told the Conservation History Association of Texas' Texas Legacy Project many years later. It was no surprise that when Sansom met her at a conference in Washington, DC, in the mid-1970s, they hit it off immediately and she encouraged him to return to Texas.

"Sharon was very involved in the environmental community in Texas, which at that time was still relatively small," Sansom recalled. "I had absolutely no idea that there were any environmentalists in Lake Jackson. She really worked on me to come back, and I think to this day that was a factor for me."

Sansom and his family did return to Texas in 1976, settling back in Lake Jackson. Sansom commuted to the Energy Institute in Houston, where as deputy director he created an Energy Extension Service responsible for community outreach and grant writing. He reveled in the media relations involved in the job—a weekly radio program called *Energy Insider* and a contributing editor position at *Houston City Magazine*, focusing on energy. He was particularly interested in passive solar technology—the use of building windows, walls, and floors to collect heat in the winter and reflect it in the summer, without requiring electrical devices. Most importantly, Sansom discovered the tiny but growing environmental community in Texas.

"It was almost like it was a guerilla underground that would appear every so often, and Andy was well accepted within that group," said attorney Jim Blackburn, who helped pioneer the practice of environmental law in Texas in the seventies. As a law student at the University of Texas at Austin starting in 1969, he found virtually no environmental law courses available. Three months after Blackburn started law school, the National Environmental Policy Act was passed, followed by other landmark legislation in subsequent years, and the foundation of modern environmental law was laid.

In law school, Blackburn made the lowest grade in one class, but he won a national competition for environmental papers with a submission about the law of the ocean. After earning a law degree, he received a master's degree in environmental science from Rice University.

As a young lawyer, Blackburn was among the first to examine how new federal environmental laws—the ones passed during the Nixon administration—applied to water and air pollution in Texas. Income from that evolving area of the law profession was meager, leading Blackburn to look for other ways to make money. Teaching seemed promising, and his Rice University connections proved helpful in landing a job there at what was then known as the Wright Center for Community Design, "a kind of environmental think tank associated with the Rice School of Architecture," Blackburn explained. "I started teaching architecture and then environmental planning, and then began teaching environmental law at Rice in the early eighties."

Another way that Blackburn scrambled to make ends meet was through consulting. One of his early consulting clients in the 1970s was The Woodlands, a planned community in North Houston being developed by the oil and gas magnate George P. Mitchell, who would go on to perfect the hydrocarbon extraction technology called fracking.

In the mid-1960s, Mitchell had allied with Terese (Terry) Tarlton Hershey, a tireless crusader for Houston parks, to oppose efforts by the US Army Corps of Engineers to line Buffalo Bayou with concrete. Garnering support from the neighborhoods along the bayou, they organized the Buffalo Bayou Preservation Association and fought to save the fifty-two-mile tidal river that meanders through the city from being straightened and covered with concrete so it would flow faster.

Hershey mobilized garden clubs, civic leaders, corporate titans such as Mitchell, and politicians such as George H. W. Bush to stop the bayou project. She went on to launch a series of other conservation groups and ensure that the nascent environmental community flourished through her strategic networking.

Hershey's endeavors likely would have led her to Blackburn one way or the other, though they actually met through Blackburn's mother-in-law. Not long afterward, Hershey introduced Blackburn to Sansom. The two men shared a personal interest in coastal issues, such as preserving habitat and species diversity, and they enjoyed the camaraderie of fighting what was considered the good fight against all odds.

"I considered our paths very much parallel during that time period," Blackburn recalled, adding, "It wasn't a place where environmental activism was rewarded."

Another member of the guerilla movement was Terry O'Rourke, a colorful attorney of Irish descent who had worked as a law clerk for a federal judge in Washington, DC, on the legal challenges to the Alyeska Pipeline across Alaska. In 1972, he moved back to Texas and the following year was hired by the Texas attorney general to be "his swinging sword on pollution prosecution," O'Rourke told the *Houston and Nature* podcast. The attorney general was John Luke Hill Jr., a Democrat who was elected in the waning days of the party's dominance in Texas before the Republican Party's ascendancy. "Hill was elected saying, 'I'm going to clean up pollution,'" O'Rourke recounted to David Todd of the Texas Legacy Project. "And he approved of my unorthodox behavior."

The brash young lawyer opened an AG's office in Harris County, where Houston is located, and started taking on local giants such as Armco Steel, Tentex Alloy, and Champion Paper over air and water pollution. "I came down with the kind of mentality of prosecution that I'd learned in Washington, DC," O'Rourke continued. "I . . . put [out] a ten-most-wanted list and . . . said, 'Here are the companies, and you're the most wanted, and I'm going to take you one at a time.' Well, they were not used to being dealt with in that way."

The biggest polluter was the City of Houston, O'Rourke told Todd. "You literally had floating turds here. It was raw sewage because we had developed faster than we had built the capacity to handle our own sewage. You could fill the Astrodome to the brim twice a day with raw sewage." It was only in 1972, with passage of the Clean Water Act, that wastewater was required to be treated to minimum standards before being discharged into regulated waters.

O'Rourke took on the Houston Ship Channel, too. The Port of Houston was booming after becoming the first in the country to introduce container shipping. "The ship channel would catch fire sometimes. It was anaerobic. There was simply no oxygen in the upper water." Being an environmentalist in Texas at that time "was a little bit like being in favor of the Communist Party," O'Rourke said.

O'Rourke and Sansom later joined forces to create a nature preserve in an unlikely spot. West of Tomball is a small, canopied forest and bird haven that belonged to a hermit by the name of Elmer Kleb. In the late 1980s, O'Rourke called Sansom—who was running the Texas Parks and Wildlife Department by that time—to ask for help in halting a forced sale of Kleb's 133 acres to pay back taxes. O'Rourke, as assistant district attorney for Harris County, was calling on behalf of a county commissioner who wanted to stop the sale due to Kleb's inability to defend himself and because of his poignant story.

Kleb had been declared mentally incompetent following decades of living as an autistic recluse who transplanted native trees and vegetation onto his property and shared his century-old house with a buzzard. The house had no electricity or running water, making for a particularly peaceful refuge for birds. Sansom arranged for TWPD to give a grant to Harris County that enabled it to buy the land, pay the tax bill, create a trust fund for Kleb, and create the Kleb Woods Nature Preserve and Center.

While environmentalists loomed large in Sansom's world in the late seventies, it was a Houston architect inspired by Frank Lloyd Wright who led Sansom down the path not yet taken. Walter Duson was his name.

Chapter 8

The House That Andy Built

Architecture had intrigued Sansom as a child growing up in a planned community with distinctive design by a Frank Lloyd Wright disciple. The midcentury modern houses set amid ample green space distinctly shaped his memories of childhood. As an adult, his fascination with building design was rekindled by pioneering architects such as Pliny Fisk, whom Sansom had met during an educational tour staged by the Federal Energy Administration, and George Way, an architect at the University of Houston Energy Institute, where Sansom worked.

Fisk was an early champion of energy efficiency in buildings and led a loose but expanding network of architects, environmentalists, educators, and others around the state who were of a similar mind. Way was an architect who wove together passive solar technology and environmental design and was a widely sought speaker on the subject. These early adopters of energy conservation clearly appealed to Sansom. However, it was Houston architect Walter Duson who really captured Sansom's imagination.

Duson grew up in El Campo, a small town near the Texas coast, like Sansom. Both were raised in Presbyterian families. Duson designed passive solar buildings, which were of particular interest to Sansom because of their use of windows and materials to collect sunlight for heating and cooling the structure. Duson used Frank Lloyd Wright principles that he had learned at the University of California in Berkeley and adapted to the Texas vernacular. His style combined light and airy design with energy efficiency.

The rapport between the two was instant when they met at an Energy Institute conference on energy-efficient building design. "Walter was one of the few practicing architects in Texas who was determined to use energy conservation in their buildings," Sansom explained. "I wasn't just excited by his work and passion; he was a kind and generous person with whom I became good friends."

In fact, Duson wanted Sansom to come work in his firm because of their shared vision. That was enough to persuade Sansom to completely change careers. He applied to the University of Houston architecture school and was making plans to attend when he discovered that he would have to start over as a freshman in undergraduate studies. It was a bridge too far, and he decided not to go to architecture school, though he did not abandon his interest in building design.

On the contrary, Sansom seized the first opportunity that came up to apply his ideas about conservation and design. Not long after he and Nona arrived in Lake Jackson in December of 1976, Sansom bought a residential lot on a bayou in a subdivision just outside of town. He couldn't get out of the rented townhouse where his family was living and into a home of his own making fast enough.

"He thought this lot would be the perfect place for a passive solar house, and he was working with people who could help him design the perfect house for the lot," Nona recounted. "At the dinner table one night, he told the kids and me that that was exactly what he was going to do—design our house to go on that lot."

Nona asked for three things in the house:

- A kitchen pantry
- A guest bathroom with direct access, not via another room
- A coat closet near the front door

Sansom's design was inspired by the likes of Duson, Way, and Fisk. For the blueprint plans, Sansom had access to some of the most advanced engineers and architects in energy conservation at the University of Houston. The architectural style featured a traditional dog run design—the same type used to cool pioneer homes in Texas—and combined passive solar technology and reused materials. About a third of the structure was built of recycled wood that Sansom scavenged in the area. He joined a local demolition company that was tearing down a cooling tower at the original Dow Chemical plant in Freeport, where he harvested redwood with a chain saw. In Beaumont, he salvaged flooring from an industrial company, and in Milwaukee, he hauled away stained glass windows and doors from a Catholic hospital.

"He spent enormous amounts of time scavenging old buildings, going back to the nineteenth century, for construction materials," Nona recalled. "Many of the most important principles for passive solar were found in farmhouses and other buildings built prior to air conditioning." Sansom cleared the site, and then a local construction crew actually built the house. For the finish-out, he helped sand and polish the floors, among other final touches.

Some of the details looked similar to a house Duson had designed for his parents on the Texas coast at Carancahua Bay—a raised roof, louvers that drew in air, and deep eaves. The bay house had no air conditioning. Sansom's project was ambitious and all-consuming.

"Weeks went by with no more mention of our house," Nona recalled. "The kids and I would ask how it was coming and what it would look like, and he would say it was coming along fine and we would see it when it was done. Weeks became months and he would say nothing more. That's when I suggested he was 'pregnant' with the house—something was cooking but was not to be seen until it was finished."

The dwelling was oriented to the south to catch prevailing breezes and featured two dog runs—large breezeways for natural cooling. It had deep eaves on the south side of the house that kept out the sun in summer and allowed it in during winter. A high roof rose about twenty-five feet above the floor and featured a large series of louvers in the top area that would draw a breeze, a "thermal chimney."

The house could function without air conditioning more often than conventional ones, though AC was still needed for high summer when temperature plus humidity along the coast can easily reach 110 degrees. In winter, conventional heating was also needed at times.

Still, there were failures. A solar hot water system never functioned as intended because it was one of the first in Texas, and "nobody knew how to do it at that time," according to Sansom. Also, mildew grew in the hallways from exposure to the hot, humid climate.

More worrisome, the house project had exerted enormous pressure on Sansom and his family. He had overseen construction while commuting to his job in Houston—a sixty-mile round trip. The project was imposing enormous strain on the family, draining finances, and worsening tensions with Nona that had already started building at the end of their time in the Washington, DC, area. While they were still living in Annandale, Nona recalled, Andy came home from the office one night. She had two young children hanging on her while she was trying to cook dinner after working her part-time job that day. She asked him to watch the kids for a few minutes while she finished cooking, and he exploded. "He picked up the pot lid and slammed it down—the dent is still there—and said, 'I work all day and all you have to do is take care of these kids.'"

Friction had been growing for a while as Sansom often traveled on business trips and worked late at the office, leaving Nona to care for the children and hold down a job as a recreation director on her own, with no family around to help. Nona felt they were not communicating well and suggested they go to marriage counseling to address their problems. But Sansom refused, saying he didn't see any problems. Nona did. She contemplated staying behind in Virginia as Sansom was making plans to move back to Texas for the University of Houston job, though she didn't do it.

"Nine months after his big announcement at the dinner table, he brought home a big roll of plans and blueprints," Nona recounted. "He was the only one of us who had had any input in the grandiose plans he spread out before us. I said 'Where's the pantry? Is there a bathroom where a guest can go without going through a bedroom?' Of course, that put him out."

Sansom offered her a choice of the coat closet or the kitchen pantry, and she chose the pantry. For the guest bath, another door was put in that allowed access from the living area instead of from the master bedroom. She got two of her three requests.

While the house was being built, Nona worked as a substitute teacher. She'd been out of the classroom for ten years and wanted to discern whether she still loved the work enough to go back to fulltime teaching. "It was a positive experience and let me know I was ready to resume my career," she recalled. Their five-year-old daughter, April, would be old enough for kindergarten by September 1978, and that's when Nona resumed her teaching career. "l loved teaching and taught the third grade for six years in Lake Jackson," she recollected.

It was an impressionable time for the eight-year-old Andrew. "He tried to have us out there at the site during construction," Andrew said of his father. "We would sit in his lap and look at the drawings that he was doing. Living through that whole experience really influenced me in terms of wanting to become an architect. That's where it started."

Sansom wanted to move the family into the house even before it was finished. The weather was cold and damp in February 1979, and there was no supplemental heat beyond solar energy. Andrew and April were young, in third grade and kindergarten, and Nona was teaching school.

"Not only was there no heat in the house, but the driveway had not been put in, and there was just gooey mud for the 175 feet between the house and road," Nona explained. "It would have been impossible for me to get the kids and me out to school every morning."

On a deeper level, they also were out of sync. "I perceived the house to be one of the principal causes of my life at that time," Sansom explained. "No one had built a modern-era passive solar building in the Houston area."

Nona said her husband's obsession with work was taking a toll on his home life.

"It wasn't worth it to him to be at home very much and to be with his kids and me," Nona recollected. "I felt a great deal of love and admiration for Andy. And I just thought he doesn't understand me, where I'm coming from. I couldn't sit down and have a rational conversation with him. And it turned into complaining. By February 1979, our family had to move out of the townhouse where we had been living and start paying mortgage payments on the unfinished house."

She decided to go on her own to a counselor, who "was just staring at me the whole time, just listening, very intently. Like nobody had listened to me in a long time. And he said, 'It sounds to me like Andy is taking care of himself. Now you need to take care of yourself.' That's all he said."

The next day, she took off from work, got a substitute teacher to cover her class, found an apartment, and signed a lease for it. "I said 'Andy, I'm sorry, I can't move into the house,'" she said. "I'm going to get an apartment and I will take the children with me." Sansom was "manic," Nona recalled, and insisted the kids could live with him in the new house.

Nona carried through. She moved into the apartment near the school the children were attending and where she was teaching the third grade.

"We didn't really talk for a long time," Nona recalled. "We would exchange kids and say, 'Hi, how are you doing?' But we didn't sit down and have any talks until Andy called."

As Sansom remembered it, "She basically said, 'Call me when the house is done and maybe I'll come back.' And I think we thought it was over." He admitted, "It was entirely my fault."

The split was a cooling-off period for both of them. Sansom moved into the unfinished house and saw the kids on the weekends. And yet, after seven or eight months, he was heartsick. He wanted to save his marriage. And this time it was Sansom who wanted marriage counseling. He knew a United Church of Christ minister he'd met while doing community relations for the Energy Institute. "It just worked out so beautifully because when I suggested a counselor before, he hadn't been interested," Nona recalled. "But when he found someone he could feel comfortable talking to, it worked."

Sansom met with the minister first, then asked Nona whether she would meet with him before they went together. "I said, 'absolutely.' Of course, I had been waiting for years, literally," Nona recounted. "What came out was that Andy thought from growing up in his family, that you just fell in love and got married. And that was it, that you kind of checked that off the list. And, the counselor said, 'Andy, a marriage is a living thing. It's something you have to devote time to, nurture it, and keep it going. It's not something that you just mark off the list.'" And then the counselor said, "Nona, when are you going to stop griping? You know, it wasn't all Andy."

Once Sansom realized how hard his absences were on the family, he started spending more time with Nona and the children. "He was more mentally and emotionally there," she recalled. "And that's been a struggle for him. My impression is that it was like an awakening for Andy. It was an epiphany."

The revelation was that Sansom's ambition was bigger than he had realized, and that single-minded drive is common among those particularly passionate about their work. "He's so dedicated to his cause, to conservation, and to doing great things," Nona observed. "And every once in a while, he'll say, 'Wait a minute, maybe I'm a little too concentrated on this side and not quite enough on the other side.'"

The estrangement opened communications lines, which had been blocked or not existent. And over the years, they went back to counselors a few times, to "get us back on track," Nona acknowledged. They reconciled, and Nona agreed to move into the new house along with the children. That didn't end Sansom's love affair with architecture, however. In fact, he had more time to pursue pet projects because his work at the Energy Institute had shrunk into a part-time job establishing an energy extension service in collaboration with Texas A&M University. He went on to design another building, the offices of the South Texas Council of Girl Scouts.

Other avocations opened up to Sansom because of his available time, inherent sociability, and Nona's steady income as a teacher. One was the restoration of the Brazoria County Courthouse as a historical museum. Sansom was spending more time in Lake Jackson (located in Brazoria County), where he wanted to be an active member of the community. Sensing Sansom's desire for civic engagement, the local newspaper publisher, Jim Nabors, recruited him to raise funds for restoring the courthouse.

Sansom had never done fundraising before, though it turned out to be a formative experience. He worked in Nabors' newsroom, where he had a cubicle just outside the boss's office. Sansom used an IBM Selectric typewriter to draft fundraising letters and then churn out copies. Nabors was quite a taskmaster. "He would come out of his office and if that typewriter wasn't cranking away, he would become upset because he recognized that we had to keep the letters going out to keep the money coming in. And I always related that to his whole orientation—which was like a printing press because we had a printing press there. So that's where I got my initial experience in fundraising."

The campaign raised several hundred thousand dollars, a considerable sum in those days. "And I learned on the job," Sansom recollected.

His most daring avocation was politics. It was one of the main reasons Sansom wanted to move back to Texas from Washington—to explore a possible run for public office. The Democratic Party had dominated Texas politics since Reconstruction, and virtually no Republicans held statewide office. Sansom's father was a loyal Democrat, yet that didn't sway the son away from his Republican leanings. He'd been impressed by the environmental credentials of leading Republicans in Washington such as Nixon and Rockefeller. In fact, Sansom sensed an opening for the GOP in Texas, believing it was time for a two-party system. "We were the insurgents," he contended.

It was the mid-1970s and his feelings of carpe diem were shared by Republicans in the state house, where they were vastly outnumbered 132–18.

The tiny caucus of mostly young and enthusiastic GOPers included a representative from San Antonio by the name of Joe Straus, a moderate Republican in the mold of the Bushes, who backed business-friendly policies that siphoned off Democratic voters. Straus eventually would become speaker of the house before being ousted in 2019 by hard-right Republicans pushing "bathroom bills" and other social policies that often overshadowed the interests of business and industry.

Another member of the close-knit caucus was Kay Bailey Hutchison, who had worked as a TV reporter before being elected to the state house. She later served as a US senator from Texas for ten years and thereafter as the US ambassador to the North Atlantic Treaty Organization in the 2017–21 Trump administration.

"We were a chummy, merry little band of Rs," Hutchison reminisced about the legislative caucus in the seventies. "We would go to camp in the summer, it was called Camp WannaMeetaGOP. We would do sports. These were really good bonds, they lasted forever."

The political pep squad was organized by Sansom and several other novice Republicans who were excited about the party's enormous upside potential in the state. Texas had been dominated by Democrats for a century. One of the organizers was Cyndi Taylor Krier, a staff member for Republican US Senator John Tower who would later become a state senator from San Antonio. Another founder was Chase Untermeyer, executive assistant to the Harris County Judge and later a state representative from the Houston area.

While some of the group members were elected officials, others like Sansom were aspiring ones, and still others—like Karl Rove—were political consultants. The friendship between Hutchison and Sansom endured as their paths occasionally crossed. In 2004, Hutchison needed some help as she was sponsoring a bill in the Senate to designate El Camino Real de los Tejas as a national trail. The trail, which was the main overland route from Louisiana through Texas to Mexico City during the Spanish colonial period from 1690 to 1821, connected a series of colonial missions and posts that enabled Spain to colonize a swath from North Louisiana to South Texas.

When Hutchison discovered that some landowners around San Marcos didn't want to grant easements for the trail, she asked Sansom to help. By that time, he was working at Texas State University in San Marcos and happy to oblige. The required easements were secured.

As Sansom was contemplating politics in the mid-1970s, his potential entrée was aided by Rove, a young political consultant in Houston who was laying the groundwork for George H. W. Bush's presidential run in 1980. Rove was the only staffer on the campaign for the first year after its launch in 1977, though the effort was chaired by Houston lawyer and ex-Democrat James Baker III.

The first volunteer on the campaign was Sansom, who had met Rove through Untermeyer, a former newspaper reporter who was collaborating with Bush on a memoir and then later became an executive assistant to Vice President Bush.

What struck Rove about Sansom was his likeness to Albert Bel Fay, a twentieth century Republican Party leader in Texas and nationally who was a fervent supporter of parks. He had crusaded for the creation of national parks on Padre Island and in the Guadalupe Mountains during his campaigns for land commissioner in Texas in 1962 and 1966. He lost both races.

Sansom met Bel Fay through the widening Republican circles of the latter seventies and got to know him better in the eighties. While Sansom was executive director of The Nature Conservancy in Texas, Bel Fay's son-in-law, Frank Smith, was chairman of the board. They would go duck hunting in Southern Louisiana each year.

"Because of [Sansom's] familiarity with Bel Fay's thinking, he was wondering whether he should run for land commissioner," Rove recalled. "He was a bright young guy and fun to be around, very smart and with a lot of integrity. But I never took the idea of him running for office seriously." Sansom lacked name familiarity and financial resources. "And he was living in Lake Jackson in a weird house," Rove continued. "That house was constantly under construction. Poor Nona and poor house."

In any case, Sansom went to work on the Bush campaign, doing research and writing issue papers, and becoming close friends with Rove as they put in long hours together at the funeral parlor that had been converted into campaign headquarters. "They had to come and take a draining table out of what became my office," Rove recounted. "We would take naps, devour some food, and hit it until one or two o'clock in the morning. But we had a lot of fun."

Rove named his only child, a son, after Sansom. Actually, his wife at the time picked the name. "She said, 'I want to have a name that whenever I hear it, I have a good reaction to it,'" Rove explained. "So she picked Andrew."

Sansom was made chair of the Bush campaign in Brazoria County, which went for Bush in the primary. Sansom then was elected as a Bush delegate to the Republican National Convention in Detroit, where he could observe the political machinery at a painfully close distance. Bush, as the more moderate candidate, lost the presidential nomination to Ronald Reagan, the firebrand.

At the same time, Rove opened doors for Sansom to work on other campaigns. In 1978, Bill Clements was running for Texas governor after serving in the Nixon administration as deputy secretary of defense. Sansom and Clements got along well, having both been in Washington at the same time, leading to Sansom chairing the Brazoria County campaign for Clements. He narrowly defeated a former Texas Supreme Court chief justice to become the first Republican governor in Texas since Reconstruction.

The third campaign that Sansom worked on was also at the behest of Rove, who had recruited Republican J. E. "Buster" Brown as a state senatorial candidate following the successful candidacy of Clements and the popularity of the Reagan/Bush ticket. Brown was from Lake Jackson, Sansom's hometown, and the rapport was good.

More than forty years later, Brown smiled fondly at the mention of his young campaign manager, who helped him win his first race and launch a decades-long political career. The two Lake Jacksonians still mingled in the same circles as the senator was an éminence grise in the water world for his landmark legislation creating a statewide water planning process for Texas.

In 1980, Brown ran against incumbent A. R. "Babe" Schwartz, a progressive Democrat and leading environmentalist in the legislature. Sansom felt "somewhat conflicted," given his longstanding support of environmental conservation. Still, he viewed the Democratic Party as the "PRI" of Texas, referring to Mexico's Partido Revolucionario Institucional, which dominated the country's politics from its founding in 1929 until the end of the twentieth century.

Sansom set aside his personal agenda but not his personal ethics.

"Campaigns against Schwartz had traditionally been anti-Semitic," Sansom noted. "That infuriated me, and I insisted I would leave if we ever got into that kind of stuff. And Buster agreed. It was a clean campaign, largely because he was a great candidate."

When Brown won, Schwartz called to congratulate him and invite him to dinner at his Galveston home. Brown asked Sansom to accompany him, and Sansom observed as Schwartz graciously offered to prepare the freshman Brown for his new office. "It was the most wonderful gesture that I can ever remember in politics," said Sansom. "They were on opposite sides of the fence, Schwartz had been in the legislature for twenty-five or so years, he takes this young whippersnapper, who's a Republican, and basically does everything possible to help him succeed in his job."

Why did Schwartz do that? "Because he was a good man. I think his main concern was for the people of his district."

As Sansom soul-searched about running for elected office himself, he realized that his firm belief in a two-party system was an underlying motivation. "More than anything else, I felt like the fact that there was essentially only one party in Texas was bad," he recollected. "The state would benefit if there were two strong political parties. My father was a lifelong Democrat, a hardcore political activist, a conservative Democrat—and he would rail about the crooked county commissioners. I'd say, 'Dad, what you need is a Republican district attorney.' And he'd go, 'Oh God, no, we can't do that.' Checks and balances—that was a great part of my motivation to consider running."

Yet, the reality of politics was dismal to Sansom. Working in the trenches of political campaigns ultimately convinced Sansom that a politician's life was not for him. "I was really unwilling to subordinate everything else to my ambition," Sansom recollected. "It looked a whole lot less romantic than I had thought from the outside. I couldn't see myself going to fundraisers every single night."

Chapter 9

Freelance Writing Doesn't Pay

Texas' first nuclear power plant was two years behind schedule and millions of dollars overbudget in 1979. The South Texas Project, as it was commonly known, was under enormous pressure to speed up construction and bring down costs. It was also under intense scrutiny in the wake of the partial meltdown of the Three Mile Island nuclear reactor in Pennsylvania in March of that year. Antinuclear sentiment had been gathering pace during the seventies, fueled by the suspicious death of nuclear power activist and whistleblower Karen Silkwood in 1974. (Silkwood was immortalized by the actor Meryl Streep in the 1983 film *Silkwood*.) Environmental groups such as the Sierra Club were agitating for greater oversight of—and even a halt to—the expansion of nuclear power that followed the 1970s energy crisis.

Even some workers at the South Texas Project, jointly owned by utilities in Houston, San Antonio, Austin, and South Texas, were worried that the construction contractor, Brown & Root, was cutting corners to make up for lost time and money. Brown & Root had no experience as a nuclear contractor when it started work on the twin-reactor plant near Bay City in 1975. A particularly concerned inspector at the project started collecting documents that allegedly showed irregularities in the construction process and wanted Sansom to know about them, having read articles Sansom wrote for *Houston City Magazine* while working for University of Houston.

"I got a call from an inspector at the South Texas Project who had a trunk full of fake documents concerning the construction of the nuclear plant, and he wanted to give them to me," Sansom recalled. "The pressure to complete the plant was so great that the inspectors would simply sign the forms and not do the inspections. For example, in the containment vessel on Unit One, there were periods in which they went for almost a year without ever actually performing an inspection. And all of that documentation was fabricated."

Sansom had time to investigate the inspector's claims, given his scaled-back duties at the Energy Institute following the departure of Jack Howell, who had originally recruited him. The offer of secret documents riveted Sansom. It had overtones of the Karen Silkwood case. Silkwood was killed in a car accident just before meeting with a journalist to give him documents allegedly proving that she had been contaminated with plutonium at her workplace. Moreover, investigative journalism had gained widespread credibility in the wake of the Watergate revelations by *Washington Post* reporters Bob Woodward and Carl Bernstein in the early seventies.

Sansom became an investigative journalist of sorts, building on his earlier experience writing about energy for *Houston City Magazine* and his radio program *Energy Insider*, which was broadcast on the National Public Radio station at the University of Houston. He spent six months following up on the inspector's allegation. He went to Bay City, where his informant lived, got the documents, and interviewed him extensively. "What we found at the South Texas Project was that construction began with only about 10 percent of the design complete," Sansom remembered. "That resulted in some major changes along the way and intense pressure on the contractor to get the project done. And it resulted in overruns that probably caused that plant to go from an early estimate of $600 or $700 million to something like $6 billion before it was complete."

Sansom traveled to surrounding towns, such as Palacios and Port Lavaca, and met in the middle of the night with pipe fitters, welders, and concrete technicians working on the project. While they certainly did not want to lose their jobs over leaked information, they wanted their concerns to see the light of day. They felt they could trust Sansom. "Because I lived down there, I had become acquainted with people who lived around the plant and were concerned about it," Sansom explained. "People who lived in tiny little towns like Matagorda or Palacios or Wadsworth knew this plant was going up six miles from where they lived, but they really don't think much about it until one day they heard about a meltdown at Three Mile Island. All of a sudden they became really concerned." As a good journalist, Sansom cultivated his sources and was able to paint a picture of what was going on inside the construction site.

He wrote the story as a contributing editor to *Houston City Magazine* and took it to the managing editor. "The editor and the publisher said, 'There's no way we're going to publish that story because we're afraid of a lawsuit.'" Sansom then pitched the story to *Texas Monthly* and got turned down there, too. Finally, he went to the *Texas Observer*, which happily published it—as the cover story. "I spent several months on that story and was ultimately paid about sixty dollars for it, so I couldn't afford to do it anymore. That ended my freelance career." The realization that freelance writing didn't pay was deeply disappointing. "One of the things I enjoyed most in my life was freelance writing."

Sansom had built up his hopes for a scoop, perhaps even a historic one, since only a couple of reporters were writing about nuclear power in Texas in the 1970s. One was a journalist at the *Austin American-Statesman*, and he was the other. Concern about nuclear power in Texas was largely confined to Austin, which only narrowly passed a referendum to keep its participation in the plant. While state media were mostly indifferent, national media were much more interested. CBS's weekly news program *60 Minutes* ultimately did a story on the South Texas Project, making it the second time Sansom's work led to a segment on the national show. CBS brought a production team to the area and interviewed many of Sansom's sources, triggering wider coverage in Texas, notably in San Antonio and Houston.

While Sansom's freelance writing was coming to an end, his main job was withering as well. Relations with the University of Houston were fraying. He occasionally stirred controversy with his writings and broadcasts about energy, and he began to see the "viciousness of internal university politics up close." Plus, he was tired of commuting sixty miles round trip to Houston every day. In 1979 he left the university by mutual agreement.

The end of the job left him adrift, as he didn't have another one lined up. In the lacuna, he gravitated toward environmental advocacy even though it wasn't paying the bills. His wife was again doing that. Sansom connected with concerned citizens in Matagorda and Brazoria counties who had formed a grassroots group to question whether the South Texas Project suffered any of the problems that had plagued the Three Mile Island reactor. The group planned a public forum in Lake Jackson, about forty-five miles east of the construction site, to inform residents about the project and discuss it. Sansom recruited a nuclear power expert from the University of Texas to conduct a public debate so citizens could hear both sides of the issues. The original venue for the forum was a college in Lake Jackson—until its leadership found out about the topic of the event.

"They cancelled the forum," Sansom recalled. "Ultimately we had to have it in a church because that was the only public place that would hold the meeting." Concerns about human and environmental safety were still less important than freedom for industry to do as it pleased in Texas in the 1970s.

A major turning point in Texas environmentalism had come ten years earlier, recalled Ken Kramer, who led the Lone Star Chapter of the Sierra Club for more than twenty years and was a seminal figure in the movement. The 1968 State Water Plan proposed a radical project to pipe water from the Mississippi River to East Texas and then on to South Texas, the Panhandle, and El Paso, because "Texas does not have enough water within its boundaries to meet all its needs beyond 1985." The environmental community viewed the pipeline idea as destructive to the environment, unnecessary, and extravagant, and believed better planning and more aggressive conservation could better meet water demands.

"That mobilized a nascent chapter of the Sierra Club," explained Kramer, noting that environmental sentiment had been gathering speed in Texas during the 1960s. In 1962, Ed Harte, who had collaborated with Sansom on whooping crane protection, led the campaign to create the Padre Island National Seashore, thus protecting the longest stretch of undeveloped barrier island in the world.

In the 1970s, the state of Texas felt it had to respond to the landmark environmental legislation passed during the Nixon administration. Kramer grasped that imperative soon after the bills passed, and his doctoral dissertation at Rice University examined how the ground-breaking laws would be implemented. Brilliant and pragmatic, Kramer represented the Sierra Club as a lobbyist, director, and volunteer for more than forty years, helping to shape environmental policies of historic significance in Texas.

Until that time, Texans largely viewed land and water as resources to be monetized, except for large and spectacular natural areas such as

Big Bend and Padre Island. Gradually, however, public opinion started shifting in the face of pollution catastrophes, toxic chemical spills, and a population explosion in the state and nationally. Exploitation and preservation of natural resources are always in tension, Kramer acknowledged, but "things had gone too far, and the environmental movement was trying to bring them back into balance."

One challenge was the amount of land in private hands—about 95 percent of the state was and still is privately owned. It wasn't always that way. When Texas entered the United States in 1845, it was allowed to keep its public lands, unlike most other states. However, the state sold off most of that land, or gave it away, to pay the former republic's debts, reward military service, and draw people to the sparsely settled area. In addition, large land grants from Mexico from before 1835 were grandfathered into the annexation. The largest remaining state lands today are in West Texas, where it was harder to find takers for the real estate.

The pervasiveness of private property naturally led to the dominance of private property rights over communal ones. The Texas pioneering spirit and deep distrust of government have strengthened private property rights over the years, especially regarding shared resources such as water and air. What government regulation did exist was largely laissez faire.

A prime example was the Texas Natural Resources Conservation Commission (TNRCC), the forerunner of today's Texas Commission on Environmental Quality, which is the second largest environmental agency in the world after the EPA. TNRCC was known as "Train Wreck" to detractors, who viewed the agency as easy on polluters and hard on those challenging the polluters. The TNRCC's Office of Public Interest Counsel, which represented the public interest when citizens contested agency decisions, was not independent. The public interest office reported to the very executive director and commissioners whose decisions it was challenging. Even worse, TNRCC's executive director was viewed by critics as supporting industry in some contested permit hearings, giving rise to questions about the agency's impartiality.

Still more hurdles were posed by water law in Texas. The legal frameworks governing groundwater and surface water are almost completely separate, often leaving natural resources such as springs to fall through the cracks. Groundwater ownership is a private property right in Texas, entitling landowners to pump their neighbors dry unless regulated by a groundwater conservation district.

Apart from the Audubon Society, which had had a presence since 1898, national environmental organizations were thin on the ground in Texas in the early 1970s, though the Sierra Club had formed a fledgling chapter in 1965.

More common than nonprofits and grassroots organizations were individual activists working on labors of love. They typically pursued projects on their own, often developing an expertise in areas that inspired them. Sharron Stewart shared Sansom's concerns about nuclear power. Ned Fritz worked to save the Big Thicket. Terry Hershey kept Buffalo Bayou in Houston from becoming a concrete ditch. By the early to mid-1970s, however, the fragmented efforts of individuals began coalescing as new environmental tools became available through landmark legislation in Washington.

Sansom was a major driver in one such initiative, the Texas Environmental Coalition, which formed in the late 1970s. Sansom was joined by Kramer, Fritz, Stewart, and Houston attorney Jim Blackburn in creating a big-tent group that aimed to find common ground on environmental issues among organizations that represented shrimpers, unions, and conservationists. "The coalition took a broad-brush approach," Kramer explained, because its membership was so diverse.

Sansom had to be careful about speaking too critically of polluters while he was working for a public university located in the oil and gas capital of the world. Even though his work at the University of Houston was tapering off and seemed destined to end soon, Sansom didn't want an acrimonious departure. "Andy was always a bit shy about being too activist because at the University of Houston, at that time, you could easily get fired for being an environmental activist," Kramer said.

The chilling effect of university employment posed something of a challenge as the coalition raised its public profile on the national stage. During the Carter administration (1977–1981), the coalition was "very much a player at the EPA level," recalled Blackburn.

The coalition hired a staff and lobbying arm to weigh in on state legislation that was springing up in the wake of the federal legislation. Key issues included hazardous waste, coastal protections, and environmental toxins. By the mid-1980s, however, the group was fracturing amid personality clashes over the use of coalition funds. "I was made president, in part, to use diplomacy to defuse the situation," Kramer recollected. Still, he couldn't hold the organization together on his own, and the coalition disbanded in the eighties, a few years after Kramer left the presidency.

For Sansom, the late seventies were also a time of trial and error. He took on a variety of consulting gigs to make a living after freelance writing proved a dead-end road. "All of us were struggling to make a living," Blackburn recalled. "He and I both were kind of in an area where we were trying to be employed doing honorable work at a time when honorable work was a little hard to find. Our paths were very much parallel during that time. Andy and I were both trying to make a living on the environmental side of issues but working with at least one foot in the business community."

Indeed, the business community was Sansom's next move. It was a departure from academic research and energy conservation, and only tangentially related to environmental protection. Still, it offered distinct benefits. It would end his work commute between Lake Jackson and Houston, allow him to work for an admired former boss, enable the protection of some wetlands, and—most importantly—provide a steady paycheck.

"Most of my career, up until that time, was characterized by always being on the edge in terms of being able to support my family," Sansom explained.

The job came about through a former boss at the Federal Energy Administration who had moved to Texas, set up a bulk liquids terminal company, and invited Sansom to work for him. Texas was bringing in large amounts of crude oil for its numerous refineries along the coast, and the terminal served as a liquid warehouse where barges would pick it up for distribution. It stored the oil and gas it received in aboveground tanks in Quintana, near Freeport. The former boss was Tom Noel, who had been in charge of developing and running the US Strategic Petroleum Reserve. Before working in Washington, DC, he had served as the chief of staff to Gen. Creighton Abrams in Vietnam.

"He was an incredible boss and one of the best mentors I've ever had," Sansom recalled. "First, I became a consultant to him because of the environmental implications of the terminal." Quintana is a tiny town where the Old Brazos River flows into the Gulf of Mexico, and the site sat in wetlands that were subject to regulation. The site was also subject to historic preservation requirements, because in the nineteenth century, a boatload of colonists brought over by Texas land agent Stephen F. Austin had washed up there.

From being a consultant, Sansom moved up to being a vice president of the company, which was a subsidiary of DSI Terminals in Houston. Sansom was always sensitive to image, and he recognized potential concerns about the terminal site in the small town. He proposed to Noel that they name the subsidiary something that would "resonate with the community in which we were located." He suggested calling it the Old River Company to hearken back to the river's name at its mouth.

The Old Brazos River also flowed through Freeport, where the sixty-six-year-old Freeport National Bank was restored and converted into corporate offices for Noel's company. Some of the renovation funding came from the parent company in Houston, though Sansom also raised funds from the community, making presentations to civic and other groups.

"I was the local face of the company," Sansom explained. "They needed that from a general public relations standpoint, because they were from DC and Washington and New York and places like that. I was the only executive from the community. What he allowed me to do was sophisticated. We restored an old cemetery and at least two historic buildings, both of which are incorporated in a county park that we helped create."

It was the only time in Sansom's life that he worked in the for-profit sector. "It was an adventure," he recollected, adding that he was fascinated by international shipping. "The idea that a ship would pull out of there and go to South Africa was enchanting to me."

Meanwhile, Sansom also relished his role in the small community of Lake Jackson. "I joined the Rotary Club, and I joined the church, and I was involved with the chamber of commerce." His marriage and family life had stabilized, and his treasured solar house was a reality.

The native son of Lake Jackson had returned and found comfort in the familiarity of smalltown life. His own son had vivid memories of village life. "Every Saturday morning, he would gather April and me—one or the other of us or both of us—and take us down to the Brazoswood Pharmacy, which was the place where he liked to have breakfast," Andrew recalled. "Everybody who was involved in politics in Brazoria County and Lake Jackson—that was the hangout. There was an egg sandwich named after him, it was called the Andy Sandwich. It had a few over-easy eggs and a lot of Tabasco sauce.

"We'd all sit in a booth," said Andrew, who was in elementary school at the time. "There would be a rotating cast of folks who would come through to chat with my dad. Or he would go around and talk politics with other folks there. That experience almost every Saturday morning of hanging out and listening to him talk to folks at the Brazoswood Pharmacy was a was a big, big part of my childhood."

Coffee shop culture remained a favorite setting for Sansom to network, socialize, and do business. Later in life, he held court over breakfast at the Magnolia Cafe in Austin, redolent of the hippie college town before the tech boom.

For Andrew as a child, another memorable pastime with his father was exploring the backwoods and bayous of Brazoria County. "He had this old pickup truck that he and his friend would use," the younger Sansom recollected. "Back then, that area was being developed left and right. Every time you'd go down the highway, you'd see a new turnoff, where there was some sign for a new subdivision. You'd go back into the woods and, of course, there wasn't anything built yet. There were just some dirt roads. He'd always be curious about what was going on back there. Were they close to any wetlands? A lot of times, we took off down these roads and got stuck—more than a few times."

Sansom was giving his son similar experiences to those he had enjoyed as a child, exploring the marshy woodlands of Brazoria County with just a hint of danger to make it fun.

"One of our favorite things to do was get out of the cabin, get in the back of the truck, and stand up behind the cabin," Andrew recounted. "You just hoped that no low branches came along. Nothing too death-defying, but still fun for young kids."

Just as the Sansom household seemed to be settling into a routine in 1980 and '81, the elder Andrew's job at the Old River Company unraveled as operations evolved. He began to do more work for the Port of Freeport. "That whole realm became interesting to me," Sansom explained. The port was putting together a succession plan for its CEO as he neared retirement and started grooming Sansom to take his place.

His career path hadn't smoothed out exactly as Sansom had envisaged when moving back to Texas. He usually knew what he wanted—personally and professionally. However, the personal and professional sometimes conflicted, especially during his mid-thirties. A major challenge was that small community life was hard to mesh with large environmental issues.

"The issue for me at U of H was mainly that I had gotten away from the things that I was really passionate about, which are the outdoors and nature conservation," Sansom told Dr. Jen Brown of Texas A&M University-Corpus Christi.

Still, Sansom recalled the 1970s fondly. "I became involved with—happily—a number of environmental projects in Texas." He worked in academia and explored passive solar design as a form of energy conservation. As a journalist, Sansom raised serious questions about the South Texas Project. As a grassroots activist leader, he helped found the Texas Environmental Coalition and got to know others such as Dede Armentrout of the Audubon Society and Ken Kramer of the Sierra Club.

He was well aware of his wife's role in enabling his restless career path. "One of the things I've always been extraordinarily grateful for is that Nona was a lifelong schoolteacher, and she always had a steady income," Sansom acknowledged. "That helped us get through some of these patches where my income was very erratic."

And then Sansom got a phone call.

Chapter 10

Bargaining for Ranches

Patrick Noonan joined The Nature Conservancy as an intern in 1970, and three years later he became president of the environmental conservation group known for using land trusts and private preserves to protect critical areas. Founded in 1951, the conservancy helped pioneer conservation easements, which grant tax and other benefits to landowners who forego development of their property to preserve threatened ecosystems. Noonan met Sansom in the early 1970s in Washington, DC, while Sansom was advocating for the Big Thicket's designation as a national preserve, and they maintained a friendship for decades.

In 1982, the Washington-based Nature Conservancy was looking for an executive director for its state chapter in Texas, which was struggling financially and operationally. The previous executive director had left abruptly amid intense pressure to come up with $3.5 million quickly to repay a loan that had been used to buy a stunning piece of land in the Texas Hill Country called Honey Creek. The Texas chapter had only three or four staff members at its headquarters in Austin, which was an office located downtown above a porn shop on Sixth Street—not an ideal image for instilling confidence in would-be donors.

The national staff in Washington had interviewed a number of job applicants and shortlisted a leading candidate for consideration by the Austin office. At that point, Sansom's name was put forward by Noonan, who had left The Nature Conservancy two years earlier to establish the American Farmland Trust. (He would go on to create The Conservation Fund in 1985 and remain one of the country's most respected conservation leaders.)

Sansom was elated to get the call asking him to apply. He had been fervently hoping to get back into environmental conservation, his real calling, but wasn't sure how to do it. The hiring decision ultimately fell to the conservancy's regional director, Tom Massengale, with concurrence from the Texas Board of Directors chair, Frank Smith, who later liked to take credit for the hire. Smith reportedly described Sansom as "one of the best hires I ever made." Smith had impeccable conservation credentials of his own, being one of the original members of the Galveston Bay Foundation. He also had chaired the Bayou Preservation Association, Harris County Flood Control Citizens Advisory Task Force, and the Rice Design Alliance. The hiring decision surprised the Austin staff, who were expecting another candidate, but Sansom was thrilled to be back in the environmental world. He was equally thrilled to get a regular paycheck again after several years of erratic income.

Despite his enthusiasm, he had enough doubts about the viability of The Nature Conservancy in Texas that he chose not to move his family to Austin right away. He commuted from Lake Jackson to Austin for two years, living at the ranch of Chris Harte, a board member for the Conservancy in Texas and son of Ed Harte, who had offered Sansom a job after his whooping crane report cost him his position at the Department of the Interior. Besides Harte and Smith, the TNC board included schoolteachers and college professors, who were "wonderful people, but they had no capacity to raise $3.5 million," according to Sansom. He wanted to make sure the nonprofit would survive before bringing his family to live with him.

The conservancy's rundown office lacked weather insulation, and water in the toilets would freeze during cold snaps. Another troubling feature for some was the so-called "Mayor of Sixth Street," who lived in the back of the office. His name was Billy, and he made his home there.

Sansom was so eager to get the job that he didn't do his "due diligence" before accepting the position. The urgent need to raise millions of dollars came as a shock, though as he learned more about Honey Creek, Sansom knew it would be a terrible disappointment to lose such a special piece of land. The area had been settled by German immigrants in the late 1840s and named for its profusion of honeybees and unusual limestone formation known as "honeycomb rock." Fed by springs, Honey Creek wends its way through bald cypress trees and Spanish moss–covered oaks to create a welcomed oasis amid the aridity and heat of the Texas Hill Country.

The natural beauty should have been enough to open pocketbooks, as a church minister gazing at the view with a group of potential donors from San Antonio told Sansom. They were, he said, "in the same business."

"When I have to put a new stained-glass window in the church, I can tell the congregation that it's going to be $100,000, and I'll have the money before the last member of the congregation shakes my hand after worship," the minister explained. "But if I tell them that I have to fix the toilets or hire the secretaries, it will take me six months to raise the money."

It was the most important lesson that Sansom ever learned about running a nonprofit. "Your big challenge is meeting the payroll, because donors are far more attracted to a wonderful conservation project," Sansom noted.

In this case, the wonderful conservation project also posed a problem. Sansom nosed around the Texas Parks and Wildlife Department to find out why it had not acquired the Honey Creek parcel of nearly 2,300 acres when it bought the adjacent one of almost 2,000 acres in 1974 to create Guadalupe River State Park. The Guadalupe River is an iconic Hill Country waterway that flows through dramatic limestone bluffs dotted with sheltering cypress and mesquite trees. The river was named in 1689 by Spanish explorer Alonso de Leon after Nuestra Señora de Guadalupe, the apparition of the Virgin Mary in Mexico's Villa de Guadalupe Hidalgo.

Sansom discovered that the department regretted not having bought Honey Creek in 1974 and still wanted the tract. To find out more, he turned to networking, a lifelong practice that he carefully cultivated. Feelers went out to colleagues at The Nature Conservancy in Washington, DC, and while they knew nothing about Honey Creek, they did know about some wetlands on the Texas coast that an oil consortium wanted to dispose of. Sansom started seeing the outlines of a potential deal where he would acquire the wetlands, package them with Honey Creek, and offer the bundle to Parks and Wildlife, which was particularly interested in coastal wetlands at that time.

Sansom was familiar with the coastal area, known as Peach Point and located in Brazoria County. It was only half an hour south of Lake Jackson, where he was still living while commuting to Austin. The tract was named for its wild peach trees by Stephen F. Austin, who was deeded it by the Mexican government in 1830. Often called the "Father of Texas," Austin was the most successful *empresario* in Mexican Texas. He considered Peach Point his only home, though he sold it to his sister, who established a cotton plantation there with her husband.

By the early 1980s, the 10,300-acre Peach Point property had been acquired by the oil consortium, which planned to build a storage terminal for oil tankers. However, a plunge in oil prices scuttled the plans, and the consortium was looking to rid itself of the land.

Sansom negotiated on behalf of the conservancy to purchase 8,580 acres of Peach Point from the consortium and then turned around and offered that property plus Honey Creek to Parks and Wildlife. "I said I'll give you $12 million worth of land for $6 million," Sansom recollected, arguing that property values had about doubled since they last changed hands. Parks and Wildlife took the deal in 1985 and created the Honey Creek State Natural Area and the Peach Point Wildlife Management Area, later renamed Justin Hurst Wildlife Management Area after a slain game warden. For The Nature Conservancy in Texas, the deal bolstered finances and credibility.

While rising prices for rural land in the early 1980s helped Sansom, so did falling ones in the middle part of the decade. Real estate values plunged along with the Texas economy after oil prices sank amid a glut of crude. Many landowners had a hard time making mortgage payments and paying taxes as their incomes from mineral rights plummeted in the wake of the oil price collapse. "Every ranch in Texas was for sale," Sansom recalled. "It was a horrible time for private landowners, but it was also a time of huge opportunity for me, because there was just such a need for them to sell."

One of the ranches hit by plummeting prices was Tres Corrales in the Rio Grande Valley, part of the historic McAllen Ranch, which was established in 1790 on royal land that had been granted by the Spanish king to residents of Tamaulipas, Mexico. José Manuel Gómez, a resident of Tamaulipas, petitioned the king for his own parcel, about an hour north of the current Mexican border. Gómez was granted the land in 1799, and his descendants still own the McAllen Ranch, one of the oldest ranches in the United States.

In the mid-1980s, the Tres Corrales section of the McAllen Ranch was owned by a petroleum landman named Hale Schaleben, an adventurous sort who had always been interested in nature. In the 1960s, he took his family, including two teenage daughters, on a photo safari to Africa for a month. The trip was lifechanging for the family, and Schaleben became a conservationist. In particular, he preserved native brush for two species of wild cats native to the Americas but rarely seen in Texas, ocelot and jaguarundi.

The wild cats had caught the eye of the US Fish and Wildlife Service, which tried for twenty years to buy the property from Schaleben. He refused and told Fish and Wildlife staffers to spend their money elsewhere, though he assured them he was taking care of the land just as they would.

In an unexpected turn of events, however, Schaleben got into debt and called the agency to say he was willing to sell part of his ranch. His only stipulation was that he needed the money in ninety days. The Fish and Wildlife Service couldn't move that fast, even though the agency desperately wanted the tract, so they called Sansom. He drove down to the Rio Grande Valley, met with Schaleben, made the deal, and borrowed a million dollars from The Nature Conservancy's Land Preservation Fund.

Schaleben agreed to the appraised value and promised to do whatever was needed to make the transaction work. He had one request—he wanted to keep access to a deer blind that he and his only grandson had built. Sansom returned to Austin, called together a gaggle of Nature Conservancy and US Fish and Wildlife Service lawyers, and explained that Schaleben was prepared to sell on the condition of preserving access to the sentimental deer blind. The lawyers said that was no problem because white-tailed deer were numerous, needed culling, and weren't an endangered species. They proposed to create a lease as part of the transaction that would give Schaleben privileges to hunt with his grandson and shoot three deer a year.

Sansom relayed the proposal to Schaleben, who said it was "perfect." All the parties signed papers and prepared for closing. About a month before the scheduled closing, Sansom got a call from one of Schaleben's daughters, who relayed the shocking news that her father had had a quadruple bypass heart surgery and was in Hermann Hospital in Houston.

Sansom wasted no time in driving straight to Houston and visiting Schaleben in his hospital room. His wife was sitting at the foot of the bed, and Schaleben had countless tubes snaking out of him. Sansom reached over and grabbed his hand and squeezed it. Schaleben opened his eyes and said, "How's our deal?" He clearly was worried that he was going to end up leaving his wife to have to handle the problem with the land. Sansom assured him that it was going to be fine, that everything was on schedule. Then Schaleben died.

Soon thereafter, a local US Fish and Wildlife Service staffer called Sansom to say the agency couldn't include the deer lease in the deal because of the "the people in Washington." National staffers claimed there were endangered species on the land. Sansom felt that the agency had decided to take advantage of Schaleben's death, something that left Sansom with more than a hint of resentment after many years. By this time, Sansom was out a million bucks, so he called Schaleben's wife and said the deal was on schedule to close. But the deer lease wasn't going to be allowed. Mrs. Schaleben didn't mince words: She felt she was being taken advantage of as an elderly widow who had little choice.

It was a bitter pill for Sansom to swallow. He wanted the deal to go through, and it did, but not at the cost of forcing a grieving widow to renege on her dying husband's last wishes. Sansom blamed federal bureaucracy and claimed that he never dealt with that US Fish and Wildlife staffer again.

Despite the emotional roller coaster, Sansom was on a high. He had discovered his calling in life. "It was like a hunt or chase," he explained. "I don't ever remember a time when I was more hooked on what I was doing than during that period. It was so consuming for me. I would go to sleep at night and dream about my deals. I had to start reading trashy mystery novels because I couldn't get the transactions off my mind."

The thrill was the "creative transaction"—bringing together buyers and sellers in any way necessary to protect the environment for future generations. He felt an excitement that had been lacking in previous jobs and brought that to bear in the wheeling and dealing. The "prey" of his hunt was protection of endangered animals, plants, land, and water—and their relationship to each other. Sansom had a natural instinct for spotting the kinds of people who shared his desire to protect living things that can't protect themselves. In Sansom's experience, these land stewards were few and far between in the 1980s.

Not only did Sansom revel in the dealmaking, he also became increasingly occupied with learning how land could be managed sustainably. At the time, most Texas landowners who needed help with ranching or farming would turn to state agencies such as the Texas State Soil and Water Conservation Board and the Texas A&M Extension Service. These agencies typically advised planting cultivated grasses that were not native to Texas or using range-management techniques that destroyed diversity of species and habitat, according to Sansom.

An alternative approach was that of Louis Bromfield, a pioneer of sustainable farming in the 1940s. An American, he learned soil conservation practices in France and India while living in those countries as a socialite and novelist. Bromfield also was influenced by Aldo Leopold, an American naturalist who believed that humans have a moral responsibility to care for the land. In the United States, Bromfield bought a piece of worn-out land in Ohio that had suffered from many of the same agricultural practices that led to the Dust Bowl of the 1930s. His main goal was restoring the soil; making money was secondary. He introduced contour plowing and terraced ditches, among other techniques, to restore the ecological balance of the land he named Malabar Farm, after the Malabar Coast of India. Bromfield's philosophy helped spur the "New Agriculture" movement of the 1940s. His social circle included Humphrey Bogart and Lauren Bacall, who married at Malabar Farm.

In Texas, most farmers and ranchers considered profitability their main goal, and if the land could be improved, all the better. Bromfield's mindset ran counter to one that viewed nature as a resource to be monetized by the owner. Private property rights have defined the Lone Star State since at least the 1700s, when Texians started pleading with the Spanish crown for grants of private land. They wanted to own property individually, rather than jointly with other recipients of large tracts of royal grants. The exploitation ethos evolved over centuries and today entails three main aspects: capitalizing assets rather than conserving them; profiteering for personal gain rather than collective good; and operating with as little government regulation as possible. This individualism heavily shaped the legal framework for minerals and groundwater rights in Texas.

The Bromfield-Leopold philosophy eventually found some converts in Texas. In the 1960s, Hugh Goodrich, an environmental activist in the Houston area, revitalized ranchland near Liverpool in the Hill Country by burning down the pervasive Ashe juniper to free up space and water for native plants. Following his lead came J. David Bamberger, the most famous of the pioneering Texas land stewards and an evangelist of the Bromfield philosophy. Bamberger was a self-made millionaire who had amassed a fortune as founder and chairman of Church's Chicken. He decided to use his wealth to restore a heavily overgrazed ranch in the Hill Country.

In 1969, the former corporate executive bought a 5,500-acre ranch that had been neglected and overgrazed by cattle, leaving bare patches of ground where wind and water caused erosion. The packed soil inhibited water infiltration and the growth of native plants. Only one natural spring was flowing when he bought the property. Bamberger spent the next fifty years restoring the exhausted land, which included removing invasive plants, replacing them with native ones, planting trees, and terracing hillsides. The earth responded in kind with undulating grasses and ten springs that came back to life. Selah, as the ranch was named by Bamberger, means "stop, pause, and reflect" in Biblical Hebrew. It became a mecca for those wanting to learn and practice how to revive land, water, plants, and animals.

Bamberger and Sansom met in the early 1980s, after Sansom had joined The Nature Conservancy and realized how urgently he needed to raise money to get the organization out of deep debt. Bamberger would be an ideal donor to TNC, in Sansom's view, as he lived its vision of "a world where the diversity of life thrives, and people act to conserve nature for its own sake and its ability to fulfill our needs and enrich our lives." Sansom's initial efforts to connect with Bamberger failed, however, as the Church's Chicken cofounder had been pulled back to the company to sort out some issues.

Around the same time, Sansom and TNC had decided that the non-profit's location in Austin was a problem, and the headquarters needed to move. Sansom felt the location above a porn shop was off-putting for potential donors, among others. "My wife and our children, when visiting me there for the first time, encountered two prostitutes at the entrance," Sansom recalled. "We never brought donors to the office." After a student intern worked for the conservancy, her supervising professor came to the office for a follow-up review and "was horrified to learn that her student had spent her summer in such a seedy location."

In addition, some donors thought The Nature Conservancy was part of the government, because it was located in the state capital. Moreover, some supporters only wanted to donate to projects in Central Texas. A more suitable location was San Antonio. For starters, a board member of the national organization, Belton Kleberg Johnson, owned a building across from the Alamo and was willing to make it available to the nonprofit. In addition, many donors in San Antonio had ranching backgrounds and supported the conservancy's work across the state.

In 1984, the TNC moved its Texas headquarters to San Antonio, and Sansom's family finally relocated there, too, having remained in Lake Jackson while he commuted for a couple of years. Nona got a job teaching second grade, and his children settled into their schools, uniting the family after Sansom's long-distance work arrangement. Serendipity also marked the move, as Sansom's daughter, April, enrolled in the same school as one of Bamberger's grandsons. The teenagers met and became sweethearts.

One day, not long after the move, Sansom was sitting in his Alamo Plaza office when in walked a man with a furry little ball that looked like a puppy hanging on his arm. In fact, it was a fruit bat. The man's name was Merlin Tuttle, a bat biologist who had founded Bat Conservation International in 1982. Tuttle asked Sansom whether he knew that the largest concentration of mammals on earth was thirty miles away. Sansom didn't know that. And whether he knew that about a month later, up to forty million bats would arrive. Sansom didn't know that either. Tuttle explained that Bracken Cave, a sinkhole in a hillside just north of San Antonio, regularly housed fifteen million Mexican free-tailed bats, and he wanted The Nature Conservancy to buy it. The cave was owned by the Marbach family, a prominent clan in Bexar County that had been ranching and farming for more than 150 years. Sansom contacted Elgin Marbach, the family patriarch, and talks ensued. Marbach even gave Sansom keys to the site.

Sansom was fascinated by the bat colony and thought Bamberger might be too. As a personal gesture, Sansom hand-delivered an invitation to Bamberger to visit the cave, home to the largest bat colony in the world. Bamberger was hooked.

While the tycoon-turned-land evangelist was a hero to many, he was anathema to others. He aroused intense, bordering on violent, suspicion and fear in landowners who saw him as a threat to their way of life. "Texas private landowners [are] historically known for resistance to outsiders, change, and anything that even hinted at a connection with government," Sansom wrote in *Seasons at Selah: The Legacy of Bamberger Ranch Preserve*. Landowners particularly feared the Endangered Species Act, which could force habitat protection and restoration. Potential fines and habitat conservation plans prompted some owners to "shoot, shovel, and shut up," according to a landowners' guide to the act written by the Texas Real Estate Center at Texas A&M University.

Property owners wanted the freedom to do as they pleased with their land, even if it harmed plants, animals, and neighbors. Moreover, they didn't want to be shamed about it. They also didn't want to be challenged on interpretations of Biblical dominion versus Biblical stewardship. Bamberger's clarion call to care for nature as a steward clashed with the more prevalent notion of man's dominion over all.

Critics called Bamberger a socialist—or a communist—when he argued that emotional returns on the land could outweigh the financial ones. At one speaking event, after addressing a group of landowners about how he restored his ranch, Bamberger was approached by a woman who touched his chin with her hand and then turned his head sideways. "I want to see where to put the bullet," she explained to him.

By the mid-1980s, Sansom was hitting his stride at The Nature Conservancy. He was wheeling and dealing for some of the most dramatic pieces of property in the state. Wealthy landowners were coming to him, and he was reaching out to them. His Republican credentials helped persuade them to take the initiative to protect wetlands, endangered species, and threatened ecosystems so the dreaded federal government wouldn't step in and do it for them. Sansom was mingling with oil and gas barons, flying in their private planes to scout out vast tracts. Scions of Texas dynasties became his friends.

Ed Harte was one of those. He and Sansom had remained friends after Sansom's report on whooping cranes cost him his job in Washington, DC. Whoopers were close to the heart of Harte, a former board chair of the National Audubon Society. The scion of a newspaper family, Harte's great-grandfather was a Washington correspondent for the *New York Tribune*, and his father, Houston Harte, was a cofounder of the Harte-Hanks newspaper empire.

Ed grew up in San Angelo, Texas, during the Great Depression, and he became a newspaper man himself, working as publisher of the *Corpus Christi Caller-Times* for twenty-five years until his retirement in 1987. The family was prominent in Texas, known for their philanthropy and interest in environmental conservation.

Around 1984, Sansom got a call from Harte, who explained that he and his brother Houston needed to donate their sixty-seven- thousand-acre ranch in West Texas to The Nature Conservancy *immediately*. The ranch, located between the Santiago and Rosillos mountains, was on the northern boundary of Big Bend National Park, which covers more than seven hundred thousand acres of desert, mountains, and springs in remote West Texas. It is the older and larger of the two national parks in Texas, the other being Guadalupe Mountains National Park. The terrain is dotted with sotol, yucca, prickly pear, and agave lechuguilla.

The Harte ranch featured an archaeologically significant mountain, Rosillo Peak, where, over a span of five thousand years, small bands of hunter-gatherers built temporary camps at the summit. When the Harte brothers decided to donate the land to a conservation group in 1984, they wanted it to go to Big Bend National Park, but they soon realized the National Park Service was not able to move within the time frame they needed for tax purposes. Ed Harte knew Pat Noonan, who informed them that it would take an act of Congress to give the ranch to the National Park Service. Since Congress was unlikely to pass legislation in a matter of months, they asked Noonan whether The Nature Conservancy could accept the donation and then work to transfer it to the park service.

That was when Sansom got the call from Ed Harte. So began a three-year odyssey to get Congressional approval to transfer the land from The Nature Conservancy to the park service. The ranch, known as the North Rosillos Mountain Ranch, was the biggest property The Nature Conservancy had ever owned in Texas.

For Sansom, the acquisition was a feather in his cap. While he had never been to Big Bend or the surrounding area before, he got to know it quickly. Ranch manager Buster Babb, the descendant of a ranching family whose land had also become part of the national park, ran the operation on a daily basis and lived with his wife in the foreman's house on the property. Sansom visited the spread often and oversaw the ranch for three years until it went into the national park system.

The multiyear effort to get the Harte ranch into Big Bend National Park was not a foregone conclusion. Some neighbors in Brewster County opposed it, raising questions about whether congressional support from the Texas delegation would be forthcoming. The opposition was based on two complaints. First, too much land already was under government control and no more should be added. And second, removal of the ranch from the tax rolls would raise taxes on others.

Anticipating pushback at the local level, the Hartes had promised to keep paying taxes on the property while it was under Nature Conservancy ownership, and they kept to their promise. At the congressional level, the Hartes argued that the prolific spring on Rosillo Peak could supply water for the "RV crowd" who were overtaxing the system in the Chisos Basin of the park.

In the end, Texas Congressman Lamar Smith—a freshman Republican from San Antonio and Austin—and others from the Texas delegation threw their weight behind the addition of the ranch to the park. Congress approved, and some fifty-seven thousand acres of the Harte ranch were incorporated into Big Bend National Park and named the North Rosillos Mountains Preserve. The remaining ten thousand acres of the ranch were sold by The Nature Conservancy to a private individual, and several years later that parcel also was acquired by the park service.

Across the state, another historic ranch lay on Matagorda Island's southern third, owned by a family that sounded like the Benedict clan in Edna Ferber's novel *Giant*. The Wynne family was a Texas dynasty of oil, land, money, power, daring, and eccentricity who owned the 11,500-acre ranch on Matagorda Island. The Wynne's cherished traditions included wearing gold rings at family reunions—rings that feature three intertwined cobras with a diamond in the back of each head. Blood relatives were called "snakes," and in-laws were "mongooses."

In 1937, Toddie Lee Wynne Sr. had cohosted President Franklin Roosevelt at the sprawling house on the island ranch, which belonged to Wynne's business partner at the time. A few years later, Wynne won full control of the working cattle ranch in a coin toss with that business partner, Clint Murchison, a Texas oil baron and political power broker. The Wynne family took over the ranch and the huge house.

In the mid-1980s, Toddie Lee Wynne Jr. contacted one of Sansom's board members at The Nature Conservancy, saying he wanted to talk about selling the Matagorda Island ranch to the conservancy. The board member, Tim Hixon, was personal friends with Wynne and understood the ranch's ownership situation, which was shared between the family and American Liberty Oil. American Liberty was the Murchison oil company that Wynne also had won in the coin toss, in addition to the ranch and sprawling house.

Matagorda Island is one of the most valuable natural wetlands in the United States, and its southern tip provides one of the winter homes for the last migrating flock of endangered whooping cranes in North

America. Accessible only by water or air, the forty-one-mile barrier island also serves as a resting spot for tens of thousands of migrating waterfowl and is home to alligators, brown pelicans, peregrine falcons, and the endangered Kemp's Ridley sea turtle. The island also supported domestic cattle, prompting Wynne Sr. to try to protect his herd from alligators by building earthen levees.

Following Wynne Jr.'s phone call, Hixon and Sansom went to Dallas to meet him and found out the main reason for his inquiry. He feared deep divisions in the family about how the property would be used and managed after his death. Negotiations began that day to sell the last privately owned land on the island to The Nature Conservancy, and in December 1986, the deal was done for $13 million. The next year, Wynne died of a heart attack at age sixty-two, seemingly prescient about the end of his life and a desire to get his affairs in order.

The deal provided for transfer of the property to the US Fish and Wildlife Service over the subsequent three years, relieving Sansom of another stint as ranch manager. US Fish and Wildlife then managed the tract as part of the Aransas National Wildlife Refuge, where marshes and bays support the whooping cranes. Sansom's deal meant that the entire island would be protected as a national wildlife refuge under the management of Texas Parks and Wildlife Department through an agreement with US Fish and Wildlife and the General Land Office of Texas. The 56,688 acres of sandy island and bayside marshes are known as the Matagorda Island Wildlife Management Area.

It was a major accomplishment for Sansom and a deeply rewarding end to an effort that had begun twenty years earlier when he wrote the report about Matagorda Island for the Interior Department that got him fired.

During much of Sansom's time at The Nature Conservancy, the organization provided a temporary shelter for vulnerable environments that needed protection—not a permanent home. The conservancy would hold the property for a while before passing it to a state or federal agency for permanent ownership. The nonprofit group in Texas simply didn't have enough money to own and manage every beautiful piece of land and water that came its way.

But that changed under Sansom.

Sansom's parents: Ernest (Sammy) Sansom, a munitions
specialist during World War II, and his wife, Imo, circa 1942.

Early childhood, late 1940s.

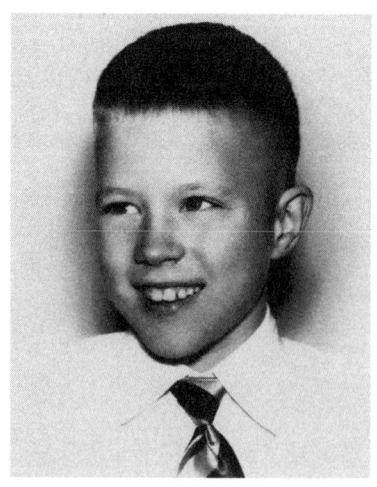

Dressed up in first tie, early 1950s.

Sighting with a bow and arrow on Jasmine Street in Lake Jackson, early 1950s.

Popular and musically talented, Sansom (kneeling) played the role of Will Parker in *Oklahoma!*, the first musical production ever performed at Brazosport High School, early 1960s.

As a budding musician, Sansom played the guitar, 1963. When younger, he also sang in the church choir and took piano lessons.

Sansom with his sister, Jeanne, who was a year-and-a-half younger, in Lake Jackson, 1960. Jeanne died in 2012.

Sansom (far left), along with fellow Explorer Scouts from the Cradle of Texas District, Bay Area Council, meeting Governor John Connally (third from left), 1963.

LJ man accepts park internship

LAKE JACKSON — Andrew H. Sansom (Andy), has accepted a one-year internship with the National Recreation and Park Association in Washington, D.C.

Eligibility requirements for internship selection depend largely upon recent graduates' personal and academic achievements in the areas of parks, recreation and conservation.

Sansom is a graduate of Texas Technological College and holds a Bachelor of Science degree in parks and recreation.

Conducted by the National Recreation and Park Association — a 25,000-member nonprofit service and educational organization dedicated to the wise use of leisure and to the conservation of natural and human resources—the internship program is designed to provide outstanding young people with post-graduate preparation in all phases of recreation and park administration.

During the year's intensive training period, Sansom will be assigned basic responsibilities in specific departments of the association and will work on projects of broad significance to the movement.

The 23-year old Texas Tech graduate was named "Young Texan of the Month" by the Texas Optimist Clubs in November, 1964 and was runner up for the "Young Texan of the Year" award in 1965. He has served as acting Park and Recreation Director for the city of Lake Jackson and plans a career in conservation and related fields, with special interests in foundation or congressional aide work.

Sansom believes his insight into legislative processes affecting the park, recreation and conservation areas has been sharpened and he looks forward to broadening his background in administrative, programming, research and development phases of his vocation in the months ahead.

"Here at Washington-based NRPA headquarters," according to Andy, "a better concept of Capitol Hill reactions to the park and recreation movement is possible."

NRPA's internship program is ultimately aimed at strengthening the park, recreation and conservation movement by increasing its number of competent personnel.

ANDREW H. SANSOM
Accepts internship

A Lake Jackson newspaper article about Sansom receiving the first internship, which was with the newly formed National Recreation and Park Association in Washington, DC, 1969.

Sansom (far left) shares a toast with his groomsmen and best man on the day of his wedding to Nona Bishop Wood, December 27, 1966.

THE WHITE HOUSE

WASHINGTON

November 26, 1975

Dear Andy:

Thanks very much for your kind letter of
November 6th concerning my appointment
as Assistant to the President.

I certainly am grateful for your words of
encouragement, and appreciate your thought-
fulness in writing.

Again, my thanks, and very best wishes,

Sincerely,

Richard B. Cheney
Assistant to the President

Mr. Andrew H. Sansom
3323 Executive Avenue
Falls Church, Virginia 22042

Note from Dick Cheney, assistant to President Gerald Ford and White House deputy
chief of staff, thanking Sansom for his support of Cheney's appointment, 1975.

Sansom presented this storyboard to President Gerald Ford as part of energy conservation campaign during the oil embargo; the storyboard was quashed due to perceived anti-Arab typecasting, mid-1970s.

Sansom (far left) with State Representative Ed Emmett at a Washington, DC, reception for Vice President George H.W. Bush and Barbara Bush, 1982.

Sansom (right front, turning towards camera) at a planning meeting during the formation of the Brazoria County Parks Board, 1981.

Sansom, far right, with staffers for Texas State Senator J. E. "Buster" Brown, early 1980s.

Sansom shows a baby alligator to children attending a Texas Parks and Wildlife Expo in the 1990s. Sansom initiated the expo in 1992.

Sansom (far left) with an unidentified person, State Representative Clyde Alexander, Governor George W. Bush, and Parks and Wildlife Commission Chair Lee Bass at the opening of the Texas Freshwater Fisheries Center in Athens, Texas, 1996.

Family photo with Nona (front) and (back, left to right) daughter April, son Andrew, and his wife Petra, circa 1995.

Sansom (right) with Lieutenant Governor Bob Bullock at the rededication of the restored Texas State Cemetery, 1997.

House of Representatives

Signing on June 30, 1995 of Resolution Ordinance
Annex of Texas Annexation to the Union

Jerry K. Johnson
District 10

Sansom (far left) and others watch as Lieutenant Governor Bob Bullock and Governor
George Bush sign a resolution related to Texas' 1845 annexation by the United States,
1995.

Governor Ann Richards and Sansom (front, left and center) rafting with two others on the Rio Grande River's Boquillas Canyon in Big Bend National Park, early 1990s.

Dear Mr. Sansom,

I was a pleasure to meet you this past weekend at the TWA convention. I am planning to be a wildlife biologist so I would definitely like to stay in touch with you. I always try to read our copy of the TP&W magazine as soon as it come in...including your column!

My family and I have been to the wildlife Expo several times and always REALLY have a good time. I have enclosed two copies of the picture of us from the convention. Could you please sign one (on the front) and send it back to me? Thanks a bunch, and again, it was very nice to meet you!

Sincerely,

Samyn Berseny

227 E. FM 772
Kingsville, TX 78363

Sansom with young girl who attended a Texas Parks Wildlife and Expo and then wrote to him about her aspirations to be a wildlife biologist, 2001.

Long-time TP & W director resigns

190

AUSTIN -- Andrew Sansom last week announced his resignation as Texas Parks and Wildlife executive director, effective Jan. 1, 2002. Sansom assumed the top post on Aug. 1, 1990, although he started with TPW on Dec. 1, 1987.

Sansom, who apparently had been quietly contemplating departure for weeks, said he was leaving TPW in good shape to pursue other opportunities, including possible university appointments or roles in Washington D.C., the private sector or in nonprofit work.

"I haven't peaked yet, but this will always be the mountaintop experience of my life," Sansom said of his TPW experience. "I'm 56 years old, and realistically I've probably got 10 years left for at least one more really important role. Nothing is forever, and it's time to pass the torch to others. My priority this fall is to make sure the public is informed about Proposition 8. For the past five years, we've done a great job of fixing things up, but if that bond

issue passes, we will be in a position to look to the future."

Proposition 8 is one of 19 constitutional amendments that Texas voters will consider Nov. 6. If approved, it would provide up to about $100 million in general obligation bond authority to repair or improve state parks, wildlife management areas and fish hatcheries statewide over the next six years.

TPW Commission Chairman Katharine Idsal said the commission would form a committee to conduct a nationwide search for a new executive director this fall.

"It's quite a blow," Idsal said of Sansom's departure. "I think it's hard for anyone to think about Texas Parks and Wildlife today without thinking of Andy Sansom. He's done one hell of a job, and I don't think anybody in Texas can argue that the state is not a whole lot better because of him."

The acquisition of Big Bend Ranch State Park in 1989, which almost 300,000 acres comprises

about half of the state park system, was an early achievement engineered by Sansom that set the tone for a tenure marked by ambitious endeavors.

Under his leadership, the state agency reduced its reliance on general tax revenue in favor of a more entrepreneurial, self-funded approach that today pervades the agency culture. He was instrumental in the creation of the private, nonprofit Parks and Wildlife Foundation of Texas, Inc. Since 1992, the Foundation has secured more than $30 million in private donations to support Texas conservation. The Foundation today leads the Lone Star Legacy campaign to fund new facilities and permanently endow all TPW state parks and other sites, an idea that could be Sansom's most lasting contribution.

Sansom was a strong advocate for private land stewardship, realizing that about 94 percent of the state landscape is in private hands. As a result, more than 12 million acres of Texas land, mostly rural farms and ranches, is now operated under some form of wildlife management plan.

Public education was a priority for Sansom. During the 1990s, he

Throughout the 1990s, Sansom battled to care for an aging state park system that needed expensive repairs. He successfully worked with several governors and legislators to secure relief, including $60 million in revenue bond authority to repair parks.

In spite of the challenge to maintain existing sites, since 1990 TPW has opened or expanded dozens of state parks and wildlife management areas, including Government Canyon near San Antonio and Stringfellow WMA (Austin's Woods) on the coast.

Sansom worked to bring the great outdoors into the big city. Realizing that urban residents will make most decisions about rural parks and wildlife habitat, he put wildlife biologists in the largest Texas cities for the first time. During his time, TPW also started KIDFISH, which stocks fish and provides rods and reels for youngsters to catch their first fish at city park ponds across Texas, as well as Becoming an OutdoorsWoman, Outdoor Kids, and the Buffalo Soldiers.

A prolific communicator, Sansom wrote two books in collaboration with photographer Wyman

A September 2, 2001, newspaper article about Sansom's resignation as the executive director of the Texas Parks and Wildlife Department, effective January 1, 2002.

Terry Hershey, an environmental conservationist who successfully battled efforts to pave Houston's Buffalo Bayou with concrete, with Sansom, undated.

Sansom (front right) with Lady Bird Johnson (in wheelchair) at a book signing for Sansom's book *Scout, the Christmas Dog*, shortly before the former First Lady passed away in 2007.

Turkey hunting in South Texas with Governor Ann Richards, who later regaled audiences with a hilarious account of killing the bird, early 1990s.

Goose hunting with Governor Ann Richards in the Texas Panhandle near Amarillo, early 1990s.

Sansom (left) deer hunting in South Texas with an unidentified fellow hunter, undated.

Sansom (left) being congratulated by his doctoral advisor, Dr. Dick Bohem, upon receiving his Ph.D. in geographic education, professor of practice from Texas State University, 2013.

Following the publication of his book *Of Texas Rivers and Texas Art*, Sansom received a Best Book of Non-Fiction Prize from the Philosophical Society of Texas. The award was presented by NATO Ambassador and former US Senator Kay Bailey Hutchinson in 2018.

Former President George W. Bush, former First Lady Laura Bush, Andrew Sansom, Nona Sansom, US Representative Liz Cheney, and Former Vice President Dick Cheney (left to right) at Dallas Fundraiser for Liz Cheney, 2022.

Landing a big catch while fresh-water fishing for steelhead in the Upper Peninsula of Michigan, 2019.

Chapter 11

You Win Some, You Lose Some

In the mid-1980s, the wind was at Sansom's back. The Nature Conservancy's profile was rising nationally, fueled by its reputation for scientific rigor and pragmatic approaches to conservation. As early as 1961, the group started using conservation easements, which protect land and water through covenants that provide tax benefits to the landowner in exchange for property restrictions. The conservancy could move much more quickly than the government if a landowner needed to act urgently.

Many of the landowners came to Sansom, directly or indirectly, through the conservancy's national office. His gift was in closing the deal, converting a lead into a transaction. Sansom understood the establishment world of money and power in Texas through his work with political elites in Washington, DC, and his social and political connections in Houston. He also understood the art of compromise, having worked the halls of the nation's Capitol. He recognized his strengths and that no small amount of luck was involved, as when preparation meets opportunity.

During the five years that Sansom worked for the conservancy in Texas, he solidified the nonprofit's role in the environmental world of Texas. He expanded the staff sixfold to nearly twenty and moved the offices from a rundown space above a porn shop in Austin to prime real estate in San Antonio's Alamo Plaza. During his tenure, the conservancy protected more than 131,000 acres, often serving as a conduit to permanent owners such as Parks and Wildlife and increasingly creating its own nature preserves. Sansom got along well with his board of directors and cultivated a network of loyal and committed donors. In short, it was the best job Sansom had ever had, up to that point.

One of his proudest accomplishments was protecting wetlands along the Texas coast, an area close to Sansom's heart. The Gulf of Mexico coastal wetlands make up more than one-third of all coastal wetlands bordering the United States. The state and federal governments both were interested in them because they protect coastal areas from storm damage and sea level rise, support commercial and recreational fisheries, and provide essential fish and shellfish habitat. In addition, wetlands serve as nesting and foraging habitat for birds and other wildlife; improve water quality by removing pollutants, nutrients, and sediments; lessen erosion of uplands; protect property and infrastructure; and support tourism, hunting, and fishing.

As recently as the mid-twentieth century, a coastal prairie and wetland system stretched nearly unbroken along nine million acres in Texas and Louisiana. Today only 2 percent of that ecosystem survives, following decades of conversion to other land uses, saltwater intrusion, insufficient freshwater inflows, and sea level rise. All these threats persist.

Given the ecological imperatives facing coastal areas, Sansom was always on the lookout for big coastal wetlands to acquire and protect. He familiarized himself with virtually every remaining wetland tract on the Texas coast and got to know the owners, be they individuals or corporations. Staying in touch with prospective land donors meant that he was in the right place at the right time when opportunity arose.

And it did. In 1986, a coastal landowner contacted the conservancy in New York about his property in Texas. Clive Runnells Jr. was a great-grandson of infamous Texas cattle rancher Shanghai Pierce (1834–1900), whose real name was Abel Head Pierce. He was known as Shanghai due to his resemblance to a banty Shanghai rooster, according to some, and because of his ruthless business dealings, according to others. Sansom connected with the descendant of the Pierce clan, which still operated in Wharton and Matagorda counties on the coast. Runnells wanted to talk about donating his property, intriguingly named Mad Island, though it's not actually an island. The parcel is situated on a peninsula that juts into Matagorda Bay and is a magnet for mosquitoes, which bit the cattle and drove them crazy, according to legend.

Runnells was interested in giving more than three thousand acres of coastal wetlands and prairie to the conservancy because of the terrain's environmental significance and commercial insignificance. The wetlands include fresh, intermediate, brackish, and saline marshes that lie within the Central Flyway for migratory birds—one of four main migration routes in North America—providing nests, food, rest, and roosts for more than 250 species. As early as 1900, the National Audubon Society focused attention on coastal birds in the area, and when it started its famous annual Christmas Bird Count, the Matagorda County-Mad Island Marsh count became one of the most prominent.

The Mad Island tract fit well with the goal of the TNC in Texas in the 1980s, which was to concentrate on coastal ecosystems. Development and overuse had wrought havoc on the delicate balance of coastal prairie, freshwater wetlands, and salt marshes. Even before being contacted by Runnells, the conservancy already had been enhancing wetlands around Mad Island, partnering with Ducks Unlimited and the US Fish and Wildlife Service to create five hundred acres of seasonally flooded areas for migrating birds, providing stopover and wintering grounds.

The talks with Runnells bore fruit. In 1987, the conservancy acquired 5,700 acres of Mad Island wetlands from the family, subsequently donating half of it to Parks and Wildlife for creation of the Mad Island Wildlife Management Area. The remainder conveyed to TPWD the following year.

The conservancy acquired another 3,150 acres in 1989, after Sansom's departure, for creation of the Runnells Family Mad Island Marsh Preserve to protect habitat for waterfowl and other wildlife. The protected areas are situated on the routes of major flyways for North American migratory birds and provide a home for a wide range of mammals, reptiles, amphibians, and crustaceans that typify the Texas coast.

Several years later, Sansom was gratified when his initial efforts on the Mad Island deal blossomed even further. The conservancy later bought another 3,900 acres from the Runnells Family with $1 million from the North American Wetlands Conservation Council and contributions from the National Fish and Wildlife Foundation, Dow Chemical, US Environmental Protection Agency, Trull Foundation, and Communities Foundation of Texas. Bringing disparate parties together like this gave Sansom great satisfaction.

"Clive's initial plan was to convey all 13,000 acres to TPWD," explained Jeff Weigel, director of strategic initiatives at The Nature Conservancy in Texas. "That happened with the first 5,700 acres. But the second deal did not get approved by the TPWD Commission due to some internal political dispute. Reportedly, Clive's response to that news was something like, 'F**k TPWD, I'll just give it to The Nature Conservancy.'"

One of the most environmentally significant deals that Sansom did during his time with The Nature Conservancy was for Clymer Meadow, one of the last vestiges of Blackland Prairie in Texas. A segment of the tallgrass prairie that stretches from Canada's Manitoba to the Gulf Coast, the Blackland Prairie runs from the Red River in the north nearly to San Antonio in the south. North American tallgrass plains featured gently rolling hills, rich soil, and undulating grasses dominated by "The Four Horsemen"—big bluestem, little bluestem, switchgrass, and Indian grass. This accommodating terrain made it attractive to settlers, especially farmers.

Today, more than 99 percent of the Great Plains has been cultivated. Agriculture, urbanization, and fragmentation have made the tallgrass prairie the most-endangered large ecosystem in North America. Clymer Meadow is perhaps the biggest and most diverse remnant of the original Blackland Prairie and among the most critically endangered landscapes in Texas. Located about sixty miles northeast of Dallas, it is named for pioneer Jim Clymer, who bought the first tract in the 1850s.

The biome attracts seasonal birds, including northern harriers that spend the winter months in Texas after summering in Canada. Large birds of prey, they got their name from the way they harry their victims. Other seasonal birds include eastern bluebirds that arrive in spring. Native only to North America, bluebirds have been cherished throughout Texas' history as signs of luck. The Navajo tribe considered the bird sacred because of its blue feathers.

In 1986, Sansom, his staff, and his board recognized that the Blackland Prairie of Texas was among the most endangered ecosystems in the country. The effort to preserve the disappearing habitat was led by then board chair Mickey Burleson, who had been instrumental in hiring Sansom a few years earlier. Burleson and her husband, Bob, had restored blackland meadows on their farm in Central Texas. Her love for nature grew out of childhood trips to the western United States, where she was fascinated by the scenery, plants, and animals.

"I could see places [in Texas] with prairie remnants that were disappearing because the people who inherited them or who bought them didn't recognize the value of what they had, and they needed to know, they needed to be informed about what they had before they destroyed it," Burleson told the Conservation History Association of Texas. Motivated by her vision, the TNC began identifying most of the remaining blackland meadows in Texas and then contacting the owners to gauge their interest in preserving the rare grasslands.

The first tract of Blackland Prairie to be protected was Clymer Meadow Preserve, a deal that materialized shortly before Sansom left the Conservancy. The small tract of three hundred acres was owned by an ageing and ill family who was unable to manage the property and needed to sell it, though they wanted it preserved.

Instrumental to the acquisition was Hixon, the conservancy board member who was a prominent San Antonio businessman. "The Nature Conservancy appeals to me because I like what they do," Hixon told David Todd of the Conservation History Association of Texas. "They buy land and preserve it, keep it going."

The Clymer Meadow deal "never would have happened without the financial support of Tim and Karen Hixon of San Antonio," Sansom said.

The grassland was attractive to Sansom and the conservancy because of its scientific and educational value rather than recreational uses. The meadow's biodiversity offered a "living laboratory" for research and teaching that were not widely available elsewhere. Two of the grassland communities, in particular, little bluestem-Indian grass and gama grass-switchgrass, were of interest due to their rarity. The landscape was distinguished by hog wallows—called "gilgai" by soil scientists—which are shallow basins, often arranged in a honeycomb pattern on heavy clay soils. Because they tend to hold water, they can host moisture-loving species such as eastern gama grass and spike rush.

Supervised field trips, tours, hiking, and bird watching could be managed to protect the terrain while research was conducted. Consequently, the preserve is regularly studied by universities, public schools, and private researchers. In the years following Sansom's departure, the conservancy stitched it together with neighboring parcels to create a 1,440-acre preserve.

For Sansom, the Clymer Preserve was more than just an environmental gem. It was a deal that catapulted the conservancy in Texas to the next level of stature—owning and managing its own land, rather than just holding land temporarily before passing it to another entity. Clymer Meadow succeeded where Honey Creek had failed. Sansom had hoped to keep Honey Creek for the conservancy as its first major preserve, alongside a few smaller ones that had materialized from bequests. But the conservancy simply couldn't afford it at the time. Clymer Meadow was more affordable because no public agencies were particularly interested in the Blackland Prairie and few, if any, private entities wanted the property.

Sansom got the conservancy back on its financial feet as he started moving it toward the creation of its own preserves, which the nonprofit would manage itself.

"Andy really changed things for The Nature Conservancy in Texas," Burleson noted. "He enabled us to get a lot of statewide publicity and to get involved in more worthwhile efforts."

While Sansom seemed to glide seamlessly from success to success, bumps did occur. One of the biggest involved Bracken Cave, where the Mexican free-tailed bats lived, and which was owned by the Marbachs, a family that traced its roots back to Germany. Bracken's scientific significance is its role as a bat nursery where females congregate each year to give birth and rear their young. Texas lawmakers later became so enamored of the creatures that the Mexican free-tailed bat was officially designated as the "state flying mammal."

The Marbach family originally owned several bat caves where guano had been mined since the Civil War. At that time, it was used for gun powder. By the time Sansom entered the picture, the family needed money to pay for the patriarch, Elgin Marbach, to go into a nursing home. Finances were tight, and selling the property seemed like the only way to raise the cash.

In the mid-1980s, Sansom began contract negotiations to buy the cave and surrounding five acres, after receiving keys to the site from Marbach. The negotiations required a property appraisal, which came in at around $25,000, far below what the family wanted and needed. The Nature Conservancy was legally bound to pay no more than appraised value and so the deal fell through. It was deeply painful for Sansom, because the property was one of the most iconic he'd worked on up to that point—and he had forged a personal relationship with the family.

Still, all was not lost—for Sansom, The Nature Conservancy, or the bats. In 1986, Tuttle moved Bat Conservation International to Austin because of its bat colony under the Congress Avenue Bridge, considered a smelly vault locally but a fragrant treasure by Tuttle.

In 1991, Bat Conservation International bought Bracken Cave and just under seven hundred surrounding acres, then added land over the years, creating a preserve that is open to members of the nonprofit and the public at managed times. In 2014, The Nature Conservancy and several government agencies partnered with Bat Conservation International to buy an adjacent 1,500 acres of land to protect the bats, endangered golden-cheeked warblers, and groundwater resources. By that time, Sansom had joined the board of directors of Bat Conservation International.

A key ingredient in Sansom's recipe for protecting nature was his ability to cultivate relationships with landowners who shared his vision and to create a way for them to preserve their environmental treasures. Many of his deals involved high profile landowners and magnificent properties. Sansom's innate discretion and social ease served him well. His accomplishments at The Nature Conservancy took it to a new level in Texas, ensuring the nonprofit a prominent seat at the table where environmental, business, and philanthropic players gathered.

And his track record didn't go unnoticed.

Chapter 12

Camo vs. Spandex

In 1970, an environmentally minded candidate for Texas Land Commissioner went hunting on a sprawling ranch just north of Big Bend National Park and was smitten with its wildness. Robert Landis Armstrong, known as Bob, gaped at some of the highest waterfalls in Texas as they slashed through canyons and at the most arid region in the state, punctuated by more than a hundred springs that dotted the desert landscape. Rare plants and animals thrived, and Native American ruins lay strewn about. Most of all, he was enchanted by the night sky, with countless twinkling stars that can only be seen in places as dark as West Texas. The ranch was captivating for a man like Armstrong who loved the outdoors—birding, fishing, canoeing, camping.

The ranch was owned by oil magnate Robert O. Anderson, who founded Atlantic Richfield and who had bought the 310,000-acre property in 1969, making him the largest private landowner in the United States. It was so remote and rugged that it was called El Despoblado, or "The Uninhabited." Bounded by the Rio Grande on the west, the ranch was a reminder of earlier times when cattle, horses, and men recognized no boundary between the US and Mexico.

The spread traced its roots back to 1905 when the Chillicothe-Saucita Ranch was established. Some of its buildings and corrals are still standing today. Over the next fifty years, the property changed hands several times, expanding along the way and eventually becoming the Diamond A Cattle Company. The operation specialized in the breeding of longhorn cattle, the iconic beeve of Texas that is cherished for resilience, intelligence, and disposition. By 1958, the ranch was about 388,000 acres—described as half the size of Rhode Island—and one of the fifteen largest ranches in the country.

After Anderson bought the property, he turned it into a vast, private hunting preserve in partnership with Walter Mischer, a Houston wheeler-dealer in banking and real estate. Though it was used for hunting, the ranch deserved more permanent protection in the eyes of Anderson, who was deeply committed to environmental sustainability. The Chihuahuan Desert property hosts eleven endangered species of plants and animals and ninety major archeological sites. Humans have inhabited the area for more than ten thousand years, drawn by the river and springs. Elevations range from the Rio Grande at 2,350 feet to Oso Mountain and Fresno Peak at more than 5,000 feet, creating a dramatic landscape.

In 1969, the same year he bought the ranch, Anderson also helped finance creation of the John Muir Institute of the Environment, named after the late nineteenth century environmentalist who is credited with creation of the US National Park Service and the Sierra Club. Four years later, Anderson founded the International Institute for Environment and Development in London, and thereafter he helped establish the World-watch Institute in Washington, which promotes sustainability of food, energy, jobs, and consumption. In a bid to protect his West Texas ranch, he started talks with Texas Governor Bill Clements in 1981 for the state to acquire the spread—either through a land swap or outright purchase. Within months, the talks collapsed amid concerns over legality, price, or both. Questions arose about whether the state could legally trade public property for private property and about the steep price tag of up to $20 million.

Anderson's passion for the environment passed to his son, Robert B. Anderson, who sat on the national board of The Nature Conservancy while Sansom was running the Texas chapter. Also sitting on the national board was Bob Armstrong, the Texas land commissioner who adored Big Bend and who had originally met Sansom during his time at the Interior Department in the 1970s. Sansom wrangled an invitation to visit the ranch during his first year at the conservancy, 1982, escorted by the younger Anderson. The tour went well, and within months, negotiations started for the conservancy to buy the tract. While Anderson boasted impeccable environmental credentials, he also was an impeccable businessman—and drove a hard bargain. Negotiations got as close as ninety cents an acre difference between the $42.20 ask and the $41.30 offer price, but they ultimately broke down. A difference of ninety cents was real money, even for a multimillionaire. Sporadic talks continued on and off through the eighties.

In 1987, the paths of Anderson, Armstrong, and Sansom crossed again at the ranch. In December of that year, Sansom was hired by Texas Parks and Wildlife to buy land for protection by the state. His work at The Nature Conservancy had caught the eye of Parks and Wildlife staff as they were looking for an expert in acquiring parkland. The agency was under pressure to expand its park holdings following a periodic review by the Texas Sunset Commission, which decides whether state agencies should continue to exist and how they should improve. Given the Sunset Commission's power and harsh criticism of the agency, leadership at Parks and Wildlife embraced the recommendations, which were approved the legislature.

The paucity of parks reflected Texans' antipathy toward government and public ownership of assets—and their reverence for private property rights. Consequently, funding for parkland acquisitions by TPWD was historically meager. Texas ranked forty-seventh out of fifty states in the size of public land holdings, and about 95 percent of all land in the state is privately held. The historical opposition to state-owned land left a vacuum of expertise at TPWD on how to acquire suitable property. Meanwhile, the pot of money earmarked for property acquisition had ballooned to $30 million.

As TPWD was trying to figure out how to carry out the legislative mandate, Sansom's name came up. He had snagged prime properties such as Honey Creek, Mad Island, and Wynne Ranch for Parks and Wildlife via The Nature Conservancy. He was hired by the agency at the end of 1987 as head of land acquisitions and quickly lived up to his reputation.

By this time, Anderson had cut the price on his ranch to about forty-two dollars an acre, below market price. Sansom wanted to capitalize on the gesture and sought moral support from Armstrong, the former land commissioner who had fallen in love with the ranch some fifteen years earlier and who was now was a Texas Parks and Wildlife commissioner.

Over the Christmas holidays of 1987, as Sansom was transitioning to TPWD, he got a call from the Anderson family. This time it was Phelps Anderson, another son of the elder Anderson, who said the family was ready to make a deal on the Big Bend ranch. Sansom relayed the message to his boss, Charles "Dickie" Travis, the head of the Texas Parks and Wildlife Department and a seasoned operative in state government.

Travis feared backlash to the acquisition from the legislature even though lawmakers had supported the purchase of more parks only months before when the Sunset Commission was breathing down their necks. "The conservative legislators were beginning to oppose acquisitions again, because it was like, 'Oh, no, well, we didn't mean *that*,'" Sansom recalled.

Recognizing the sensitivity of the deal, Travis decided that Sansom should undertake a stealth mission to visit the Andersons at their headquarters in Roswell, New Mexico. Travis devised a cunning plan. Sansom would fly up to the Panhandle for previously scheduled meetings with farmers. After he got to Amarillo, he would reschedule his appointments and rent a car to drive to Roswell. Sansom carried out the plan and got the Andersons under contract before any news leaked.

"We didn't tell a single soul," Sansom recounted. When word did get out, some pushback ensued. "To some extent, that backlash still lingers today because we bought so much land. If you look at it in the larger context, though, it's not enough."

In July 1988, the Texas Parks and Wildlife Commission formally approved the purchase for $8.8 million, with Armstrong making the motion and capping an eighteen-year mission to protect the ranch and its twinkling stars. With the acquisition, parkland in Texas nearly doubled from 263,635 acres to 492,226 acres. Big Bend Ranch State Park opened on a limited basis in 1991 and fully to the public in 2007. It became the largest park owned by Parks and Wildlife, and Sansom was thrilled because it provided protection for one of the most pristine examples of West Texas plants, animals, and springs. He was equally thrilled that the protection would benefit future generations.

Not everyone was impressed. "The purchase is the easy part, it's the long-term stewardship that matters," said David Riskind, an ecologist and expert in vegetation and restoration in the American Southwest who worked for TPWD for forty-eight years. "It was squandered stewardship, and it was Andy's fault. He was there and he didn't push far enough."

Sansom was negligent when he became CEO of the agency in not insisting on the removal of a "pet herd" of cattle that was allowed to roam the ranch by the property manager, Riskind contended. The manager was the ex-foreman of the Diamond A Cattle Company and the cattle "compromised" the sensitive desert habitat, including natural springs and native plants, according to Riskind, an expert in natural landscapes of the Chihuahuan Desert. Sansom should have brought in a professional conservationist for the property, given the size and complexity of the ecosystems.

In addition, he should have required the "toy herd" to be removed, as directed by the TPWD Commission. Riskind hinted that Sansom had leeway with his board because he developed particularly close relationships with members during the considerable amounts of time spent in executive session to review and approve real estate transactions.

"It's a cancer on the landscape," Riskind said of Big Bend Ranch State Park, because "the ecosystem didn't get a chance to meet its full potential. It was a world-class project, and it has metastasized. I was disappointed." Even though Sansom's handling of the park was "professionally hurtful" to Riskind, he and Sansom continued to work together as professionals, simply compartmentalizing their differences.

In any case, Sansom's first acquisition for TPWD was monumental—Big Bend Ranch State Park almost doubled the size of Texas' parklands. It was the biggest acquisition of his career.

Even before the Big Bend Ranch deal closed, Sansom was working on his next conquest. His spending budget was high and land prices were low. Real estate prices across the state had been decimated by the 1980s savings and loan scandal. "For Sale" signs were out, and property was available.

For some time, he had been eyeing one of the most pristine rivers in the United States—the Devils River, located in West Texas, north of Del Rio. In 1988, the moment arrived. Sansom learned from a real estate broker that more than twenty thousand acres, including waterfalls where Dolan Creek empties into Devils River, had come on the market.

Devils River is the most unspoiled river in Texas. The spring-fed watercourse dips underground where it is scrubbed clean by gravel and sand before resurfacing some twenty miles downstream. Its remote location, nearly an hour from the nearest city, protects it from pollution and preserves the purity. The channel cuts through canyons to create what rafters call a pool-and-drop river, calm stretches broken by several class two and three rapids.

The rough terrain gave rise to the river's name. Originally it was called Dacate by the Native Americans living there. After Spanish explorers came in 1675, it was called San Pedro. In 1848, Texas Ranger Capt. Jack Hays came upon the river as he and his party were exploring a good route from San Antonio to El Paso. Famous for fighting native tribes and Mexicans, he surveyed the river and environs and said, "Saint Pete, hell. This is the Devil's River," according to the Devils River Conservancy. The possessive apostrophe fell away over time.

When the real estate broker called Sansom about the river property, it was explained that a third-generation owner wanted to sell three tracts totaling 20,350 acres. The E. K. Fawcett family had once owned as many as 70,000 acres near the Devils River juncture with Dolan Creek after the family patriarch discovered the magnificent spot in 1883. Erasmus Keys Fawcett was a seventeen-year-old orphan who had signed on to help run a herd of sheep to West Texas. But when he saw Dolan Creek where it feeds into Devils River, he decided to stay, living in a nearby cave for a while before building a cabin above the creek mouth. He started raising sheep and later cattle. The Fawcetts thrived.

But by 1988, the family could no longer afford to keep the property and had to sell it. Parks and Wildlife acquired what became known as the Devils River State Natural Area Del Norte Unit, and at the sale closing, the family wept over the loss of their treasured land. About twenty years later, after Sansom had left TPWD, he did something he'd promised himself he would never do. He weighed in on an agency matter.

TPWD considered swapping the Devils River State Natural Area as partial payment for a privately owned ranch downriver that would become a new state park or natural area. Sansom publicly opposed the deal—as did outdoor advocates, environmental groups, and river enthusiasts. "It was a grand plan all right, except for one thing," wrote TPWD Executive Director Carter Smith in *Texas Parks & Wildlife Magazine.* "Very few people liked the deal, and in fact, many downright hated it." The agency scrapped the swap idea and bought the downriver property outright.

Sansom's next acquisition signaled clearly where he wanted Parks and Wildlife to go. He bought the agency's first nongame wildlife management area in 1989. The roughly two hundred acres along Galveston and Trinity bays featured oak mottes and coastal prairie vegetation—and not much else. Yet it was significant habitat for resident wildlife and migrating birds. Protected species such as diamondback terrapins also make their homes there.

The patch of classic Texas coastal habitat was named the Candy Cain Abshier Wildlife Management Area after a TWPD employee who worked to conserve wetlands and preserve historic sites. The brackish wetlands and mottes of oak and yaupon trees are a popular point on the Great Texas Coastal Birding Trail for watching migratory and wintering birds. A hawk-watch tower and observation platform overlooking the bay enable the annual Smith Point hawk watch from mid-August to mid-November.

Perhaps the most sentimental acquisition Sansom made was an abandoned railroad tunnel near Fredericksburg that had become a bat cave with a large colony. He had first learned of the nine hundred-foot-long tunnel as a mentee of Professor Bill Kitchen at Texas Tech, who had assigned him to identify railroad tunnels in the state. The Fredericksburg and Northern Railway Tunnel was one of them, having been decommissioned in 1942 after serving the Fredericksburg–Comfort line. The bat-inhabited tunnel is situated in a manmade valley where live oak, escarpment black cherry, and black walnut trees provide habitat for yellow-billed cuckoos, Carolina wrens, and Carolina chickadees.

In 1991, some twenty years after first learning of the tunnel as a student at Texas Tech, Sansom had a chance to buy it, thereby providing a home for three million Mexican free-tailed bats and three thousand cave myotis bats. That put it among the top five bat colonies in Texas, which itself leads the United States in the number of the furry flying mammals. It wasn't made a state park until 2012, and at sixteen acres, it is the smallest state park.

Looking back on his time as land acquisition coordinator for TPWD, Sansom recalled one of the most eccentric landowners with whom he dealt. A rancher by the name of Topper Frank owned property next to the Sierra Diablo Wildlife Management Area in southwest Texas, about fifty miles from the Mexican border. In fact, the ranch provided the main access to the wildlife management area, which was a critical site for the reintroduction of desert bighorn sheep to the region.

One of the agency's employees "used very poor judgment and cut the lock on the gate" between the wildlife management area and the ranch,

Sansom recalled. In retribution, "Mr. Frank enjoined the department from entering the wildlife management area. I was delegated to go out to the ranch and try to negotiate an agreement with Mr. Frank to allow us to use the road to get back into the preserve.

"We sat at his kitchen table for several hours during which time he read me the riot act," Sansom continued. "Finally, he asked if I would like to take a walk and tour the ranch headquarters, and I very gladly accepted. His wife accompanied us as we walked around while he pointed out corrals, barns, and other buildings. Coming to a closed shed, he opened the door and a full-grown female mountain lion jumped on me. About the time I hit the ground I realized she was purring and had been declawed. His wife photographed the whole thing, which I concluded was the most extreme negotiating tactic anyone had ever used on me."

After several years of hectic land dealing at TPWD, Sansom knew he had ruffled some feathers. "I wish I could have bought more, but by the time I finished, there had begun a kind of backlash," Sansom explained. Some blowback came from within the agency. A few staff members felt Sansom had gotten to the point where he was wheeling and dealing for the adrenaline high more than for preserving unique properties. Some pushback also came from lawmakers, who opposed any more land going into state hands. For the time being, however, Sansom's star continued to rise.

In 1990, Executive Director Dickie Travis announced his retirement. He had run the agency as a kind of hunting and fishing club for men, in the eyes of many. Hunters and fishermen happily bought their TPWD permits and enjoyed their stalking and angling in Texas' wide-open spaces, which were managed for their benefit.

Those with political or financial influence took it a step further. They would ask Travis to stock their lands with big game, their ranches with quail, and their ponds with bass. The ostensible reason was wildlife restoration, though the biologists were being asked to put animals into environments where they were unlikely to survive. Travis was considered amiable and shrewd and obliged the requests for what became known as "political stocking."

"A very influential family around Stephenville decided that they wanted some antelope [pronghorn]," Sansom recounted. "So, they contacted Parks and the biologist said that's not habitat for antelope—antelope cannot live there. The biologist was told that's all right, you need to get him some antelope from New Mexico." The antelope subsequently died. Agency biologists saw it as corruption.

"It was just a way of doing favors for influential people, and the biologists hated it," Sansom recalled. "And they finally put a stop to it by leaking it to the press. And that's what caused the leadership in Parks and Wildlife to change and me to be appointed."

The selection committee for Travis's replacement asked Sansom to apply for the position, and he did. During his job interview, he was questioned about what he thought his biggest challenge would be. He gave the right answer—time management. He was hired.

Going into the job, Sansom was clear in his own mind about the broad outlines of his priorities, if not the details of implementation. He wanted to put parks on a more equal footing with wildlife, diversify park visitors, make the agency more financially independent, and introduce more entrepreneurial thinking at the agency. He knew his skill at finding common ground would be essential. Even more important would be his ability to remain trusted by all sides, to retain his integrity while balancing competing interests at the agency.

Before pushing his priorities aggressively, he focused on fighting corruption, with cooperation from the existing staff. Some friends of the agency, however, didn't get the memo. About three months into his job as executive director, Sansom got a call from the speaker of the house, Gib Lewis, who said he needed some deer for his ranch near Lampasas. Sansom was stunned, given the widespread media coverage of the alleged corruption and given that Lewis was already under indictment for violation of Texas' financial disclosure law.

Sansom told the speaker that he would be put on the waiting list, and the speaker acknowledged that asking was all he could do. Not long afterward, Sansom saw Lewis at a social event during the legislative session. Lewis approached Sansom, who was standing with some of his Parks and Wildlife commissioners, and asked where his deer were. The deer never arrived, though Sansom and Lewis remained friends.

The agency's image as a good old boys' club, especially for powerful elites, was a serious hurdle for Sansom. Many landowners didn't trust Parks and Wildlife because of perceived favoritism, and Sansom realized he had to restore their confidence for anything to get done with those constituents. A major obstacle was history—hunting and fishing were the agency's original purview. In 1895, the state legislature created the Fish and Oyster Commission, and twelve years later the Game Department was added. It wasn't until 1923 that a separate State Parks Board was established. In addition, the "hook and bullet" programs were valued more highly because they generated revenue through fees for permits and licenses. The fish and wildlife budgets came almost entirely from license purchases. In contrast, birding and nature walking generated virtually nothing for the agency.

When Sansom became executive director in August of 1990, Ann Richards was campaigning for the governorship of Texas. The bouffant-coifed Democrat eked out a narrow win in November over oil and gas man Clayton Williams.

An outspoken and witty politician, Richards was only the second female governor of Texas and the most liberal one in more than twenty-five years. As a social progressive and fiscal conservative, she was intrigued by Sansom, according to her biographer Jan Reid. She considered Sansom a Republican in the tradition of Teddy Roosevelt and was impressed that Sansom had pictures of himself and President Bush hanging on his office walls.

As is common among elected officials, Richards appointed some of her biggest donors to state commissions. Her first appointment to the Texas Parks and Wildlife Commission was Walter Umphrey, the wealthy plaintiff lawyer who later investigated Sansom's car after his accident in 2005. Umphrey had made a fortune for himself and the state of Texas by negotiating a lawsuit settlement with tobacco companies. Within a week of her election, he was already saying publicly that he would be given the chairmanship of the Parks and Wildlife Commission.

Soon, newspaper headlines began appearing on page one of various publications saying that Umphrey had game hunting violations, creating enough controversy that Sansom was summoned to Richards's office. She met him at the reception desk just outside her office and escorted him into her sanctum, with no aides present. "She says, 'My daddy taught me to catch catfish on a trotline on the Brazos when I was twelve years old. And I like nothing better than to catch redfish. Everything I heard about you, you're doing everything I would want you to do over there,'" Sansom recounted. "'But it's come to my attention that some of the staff,' which I took to mean me, 'are making trouble for one of my appointees.'"

The assumption was that TPWD staff had leaked the details on Umphrey. "She said, 'Here's the deal: Appointing the commissioners is my responsibility. You running that department is your responsibility. Here's the way this is gonna work. I'll stay out of your business. And you stay out of mine.'"

Sansom quickly called an agencywide meeting of all employees and told them the governor was happy with what they were doing, except for one thing. Employees should not be inserting themselves into the process of appointing commissioners. Then he laid it on the line: "I'm not going to tolerate anybody commenting on potential appointees."

A few months later, Richards made another appointment to the TPWD's nine-member commission—only the second woman to sit on the body. It was Terry Hershey, the parks activist who had joined forces with George Mitchell in the mid-1960s to form the Buffalo Bayou Preservation Association. Hershey also was a philanthropist, having married a distant relative of the wealthy Hershey chocolate family. She supported environmental conservation causes and was urged by the Sierra Club to approach Governor Richards, an acquaintance, about the Parks and Wildlife Commission seat.

At Hershey's first commission meeting, the agenda included approval of game and fish regulations, following a long and graphic presentation of hunting and fishing harvests. Staff members would show graphic photos of fifty dead deer hanging in a processing house to demonstrate a good harvest. At the end of the meeting, the chairman allowed commissioners to speak, and Hershey was the last one. When it was her turn, she said, "I think I'm going to throw up."

Sitting in the front row of spectators were reporters for various news outlets, who jumped to their feet and raced out to the payphones in the hall to file their stories. The next day the newspaper headlines read, "Richards Appoints Anti-Hunting Commissioner to Parks."

Flak was flying. Governor Richards called Sansom for the second time within weeks. He picked up the phone and wished the governor good morning, at which point she seemed agitated and blurted out, "Andy, what are we gonna do about Terry Hershey?" Sansom replied calmly that he thought appointing commissioners was her business and running the department was his. She laughed, apparently appreciating the riposte. At the next commissioners' meeting, one of the reporters gave Hershey a barf bag from an airplane.

Sansom, for his part, took it upon himself to tutor Hershey on the agency, realizing she could be a powerful ally in advancing his goal of lifting parks programs to a more equal footing with hunting/fishing ones. While Hershey viewed the agency as a "good old white boy hunt club," she was a quick study and learned how to navigate the clubbiness. She knew how to advance issues and wield influence, having founded a number of organizations dedicated to conservation and environmental protection. She was close friends with Lady Bird Johnson and helped found the Lady Bird Johnson Wildflower Center in Austin.

When Sansom took over as executive director of TPWD, his publicly stated goal was to "make wildlife habitat and parks as important to Texas as schools, roads, and hospitals." He said he wanted to be more visible, especially to city dwellers who viewed the agency as little more than a dispenser of hunting and fishing licenses. Sansom lived up to that pledge, aided by favorable media coverage that helped him advance his agenda and defend difficult policy decisions. He nurtured relationships with editors and reporters that paid off in "good press," having learned in Washington, DC, about the power of the media.

In an effort to create more balance between parks and wildlife, he introduced the Texas Conservation Passport—a twenty-five dollar annual pass offering free entry to state parks and guided tours with the intent of attracting more visitors. Another initiative was the Texas Parks and Wildlife Expo, which was hatched after the "hook-and-bullet guys" on his staff wanted to start a big fishing and hunting event. Sansom's reply was that the "other guys" had to be involved as well. "It was a war," Sansom recalled. The TWPD commissioners took sides, too, saying, "No, we don't want those people."

"I just did it anyway," Sansom said, admitting that some very ugly staff meetings followed. The intent was not only to insert parks into the event, but also to defuse battles between park users—mountain bikers competing with equestrians for access to trails; rock climbers opposing archeological preservationists in West Texas; newcomers in crowded parks breaching camping etiquette.

The fishing and hunting event idea became the Texas Parks and Wildlife Expo—an annual exhibition where families could try out fishing, shooting, birding, photography, camping, climbing, and mountain biking with gear and guidance. It was intended to blend what Sansom called the "spandex" folks and the "camo" crowd. In park industry parlance, users are called "consumptive" if they kill wildlife such as in hunting, fishing, and trapping. "Non-consumptive" users observe wildlife as in hiking, birdwatching, sketching, and photography. The expo drew up to thirty-five thousand visitors from across the state, and over time it evolved into smaller TPWD exhibitions that were held at other larger events such as the Houston Livestock Show and Rodeo.

A particular pursuit of Sansom's was getting children outdoors. He believed they were the key to the marriage between the camo and spandex crowds. Kids were growing up in urban areas without enough opportunities to get outdoors to hike, camp, hunt, or fish. Later in his tenure, after he'd amassed more power, he introduced the Kidfish program, which stocks city park ponds with fish and provides rods and reels for children to catch their first fish.

Opening nature up to children became a lifelong mission for Sansom. He befriended a church in East Austin in 1992 and began organizing trips to take kids from the congregation hunting, fishing, and canoeing. Some of the youngsters had never before been in a canoe or skipped rocks on a river. Over time, the church bought a camp where the congregation's youth could continue to enjoy the outdoors.

From early on in Sansom's tenure at TWPD, he also sought to promote women and people of color at the agency. In 1991, he appointed the first Black woman to be division director in parks.

He encouraged the acquisition of more land where wildlife would be managed for activities such as birding, rather than hunting. In rapid fire, several wildlife management areas were acquired—Big Lake Bottom in 1990, Alazan Bayou in 1991, and Caddo Lake in 1992. Sansom promoted an outdoors license that had been introduced a few years earlier that allowed more than just hunting on public lands. The "Annual Public Hunting Permit" allowed nature watching, camping, fishing, and other activities intended to appeal to families.

In addition, Sansom opened more wildlife management areas to the public. Historically, public access to these areas was restricted to wildlife research, educational activities, and resource management demonstrations. Recreational activities such as public hunting, hiking, camping, and bird watching were allowed only if they were compatible with conservation practices and after research and educational goals had been met.

Birding especially was ripe for promotion as Texas has more bird species than any other state except California. Capitalizing on that opportunity, TPWD spearheaded creation of the first designated birding trail in the nation. The Great Texas Coastal Birding Trail opened in 1994 with ninety-five sites along the central coast, intended to promote both conservation and ecotourism. The trail eventually grew to include hiking and driving trails, birding sites and sanctuaries, and nature preserves along the entire Texas coast.

In addition to the trail, an event called the Great Texas Coastal Birding Classic was established. Billed as the "biggest, longest, and wildest birdwatching tournament in the US," the event brought together teams of birders who competed for the highest number of species spotted over three days along the Texas coast.

Some critics opposed Sansom's push for more park visitation. A few members of the National Audubon Society objected to more foot traffic and diversified uses, such as hunting. They feared that hunting would scare off animals using the parks as refuges for the winter and that more visitors would lead to "tameness" of the wildlife.

While Sansom was energetically implementing his agenda, obstacles naturally arose. Governor George W. Bush didn't want to buy more parkland until an agency maintenance backlog of millions of dollars was addressed. Agency morale needed constant care, prompting Sansom to improve salaries for biologists and recreation specialists. He decentralized the agency to save funds and delegated authority after the wildlife-stocking scandals. Within the fisheries division, he made habitat conservation a priority to help protect the investment of hatcheries.

Yet the single biggest challenge was clear—money.

Chapter 13

Splatter Zone

From the outset of his time as executive director of the Texas Parks and Wildlife Department, Sansom could see that money was going to be hard to find. Public funds were sure to fall short of what he needed to carry out his vision.

Texas historically ranked at the bottom of states in funding for parks. In 1988, Texas spent less than 40 percent of the national average on public parks—$1.79 per person versus the national average of $4.81. Since public monies were going to be scarce, Sansom started thinking early on about turning to the private sector. Extracting money from the users of Parks and Wildlife assets would be easier than extracting it from the funders of the agency. He could tap into Texas pride, historically a wellspring of dollars.

"A lot of the ideas that he had when he came into Parks and Wildlife were entrepreneurial," recalled Jeff Francell, who was hired by Sansom in 2000 to conduct land acquisitions for the agency. "He felt that state parks should have business plans that would help pay for themselves and that the agency should be more self-supporting financially."

In the first year of his directorship at TPWD, Sansom formed the Texas Parks and Wildlife Foundation, a charitable nonprofit that solicits private donations for agency projects that exceed the budget. The first project was an artificial reef made of old oil rigs, partially funded by the Shell Oil Company. The foundation won moral support from Commission Chairman Emeritus Perry R. Bass, a wealthy oilman of the old ruling class in Texas, and seed funding of $2 million was provided by Anheuser Busch. In 1996, the foundation established the annual Lone Star Steward Award to recognize private landowners for excellence in habitat management and wildlife conservation on their property.

Eventually the foundation created an endowment program to fill even more gaps left by public funding. Called the Lone Star Legacy Program, it was inspired by a Texas A&M University study that had proposed an endowment based on private donations matched by public monies. Only the interest of the principal would be spent to ensure the endowment's permanence. The Lone Star Legacy Program was the first of its kind in the nation and enjoyed enthusiastic backing from Governor George W. Bush.

Among the first five projects that received endowment funds was The Meadows Center for Water and the Environment—then called the Texas Rivers Center and shortly thereafter the River Systems Institute—at Texas State University (where Sansom would land several years later). Knowing that state recognition of private citizens builds goodwill, Sansom also created the Lone Star Legends awards to recognize Parks and Wildlife volunteers. This trio of initiatives—the Lone Star Legacy program, Lone Star Land Stewards, and Lone Star Legends—reflected Sansom's ability to build support for his causes.

While private funds clearly were helpful, it was the public ones that remained the lifeblood of the agency. Nevertheless, Sansom was determined to lessen TWPD's dependence on general revenue appropriations and increase its financial independence so he would have more ability to carry out his agenda. Toward that end, he established new revenue streams, such as an enlarged customer base for park fees and state park stores where merchandise was sold. Park entrance fees changed so that individual visitors paid, rather than vehicles. Computerized licensing programs and a centralized reservation system boosted revenue and increased efficiency of operations.

In 1990, a few months into his executive directorship, Sansom was feeling confident. It was Christmas time, and he and Nona attended her school's faculty party where he met a kindergarten teacher who was very excited about his good fortune in drawing a lottery permit to raft the Grand Canyon's Colorado River. The two men quickly hit it off and the teacher invited Sansom to join the three-week Grand Canyon trip. Sansom viewed it as a once-in-a-lifetime opportunity, knowing it could take seven to ten years to win the lottery permit. He realized the risks in being away from the office with virtually no communication lines for that long. But he was feeling good about his job and the level of support from his staff. By this time, he had managed to appoint most of the top people and felt they could be counted on.

After the Christmas holidays, he went into a staff meeting and announced that he was planning to raft the Colorado River at the Grand Canyon that summer and intended for the agency to have its proposed budget in order by the time he left. It was a tall order given that Governor Richards had mandated a 2 percent budget cut for all state agencies.

Even before the proposed budget cut, TWPD's revenue fell short of what Sansom wanted for his pet programs. Park maintenance was being deferred in the face of growing repair bills for the vast assets owned by TPWD. Infrastructure was crumbling even as the number of visitors had jumped by more than a third over the previous twenty years. Water and sewer systems were failing in more than 80 percent of parks, triggering violations from the state environmental regulatory agency.

Parks and Wildlife still had some funds left from $75 million in bonds that had been issued for land acquisition a few years earlier, and federal grants provided some income. A major source of revenue, though, was a cigarette tax of two pennies per pack, with one cent going to state parks and one cent to local parks. Undaunted by Richards' mandated cuts, Sansom saw a way to get more money. He persuaded a state lawmaker to sponsor a bill that would raise the cigarette tax to four pennies a pack to offset the drop in revenue as smoking declined.

On the first day of the legislative session in 1991, Sansom checked the Senate appropriations bill for his budget and was stunned. The cigarette tax item was gone completely—the line was zero. About a quarter of his operating budget—$27 million dollars—had vanished.

"There's only one person to talk to, " a friendly legislator told Sansom, and that was Lieutenant Governor Bob Bullock, a political legend in Texas. His bombast, intimidation tactics, and forty years in state government made him a larger-than-life figure. He was "abusive, abominable, and all-around insufferable," according to his biographers, Dave McNeely and Jim Henderson. He practiced "zero tolerance for indecision and inaction," the biographers wrote. Bullock was "the most atrocious human being who ever lived," in the words of fellow Democrat A. R. "Babe" Schwartz. Bullock himself did little to discourage such characterizations.

In spite of, or due to, his bullying, he pushed through the legislature a bill creating the state's first water conservation and management plan, equalized school funding between poor and rich districts, pushed election law reform, initiated periodic evaluations of state agencies, and established an equal employment opportunity program.

Bullock was the kingpin whom Sansom needed to consult. He made an appointment and arrived in the lieutenant governor's office, taking a seat in a chair widely known as "the Splatter Zone," directly in front of Bullock's desk. The smug lieutenant governor put his boots on the desk and declared, "I shot 325 quail in one day." Sansom later learned that was an intimidation tactic. Bullock chain-smoked and blew a puff of smoke in Sansom's face, asking what he wanted. "Governor, I'm here to see what I can do to get the cigarette tax restored," Sansom said.

"We're not spending another dime of that money on parks," Bullock declared. "We're going to spend it on cancer research." Sansom kept his wits about him and answered, "Well, that makes sense, but we can't just go cold turkey." The governor blew another cloud of smoke in Sansom's direction and said, "You're not listening to me. No more cigarette money for parks." Sansom asked what his alternatives were, and Bullock sent another puff of smoke his way. "Suicide," he said.

The meeting ended and Sansom left, unsure of exactly what had happened. The next day, a legislative hearing was held, with about fifty people signed up in support of restoring the cigarette tax. The committee chair gaveled the meeting to order and said, "Issue number one, cigarette tax, restored."

Not long afterward, the long-awaited Grand Canyon trip finally rolled round. The eleven-member group set off on the 225-mile trip that included more than seventy rated rapids and settled into a routine that included meal preparation and cleanup. One evening, Sansom was carrying boiling water for dishwashing and tripped, spilling scalding water on his foot. A blister swelled up so large and painful that Sansom found his way to the only nursing station in the Grand Canyon at that time, located at the Phantom Ranch Lodge on the floor of the gorge. The lodge had the only telephone below the canyon rim, and Sansom called Nona, having been away for a week. He got the answering machine, so he immediately called his father. "You've got to come home," his father blurted out. "Bullock wants you fired."

In Sansom's absence, his director of state parks had called a news conference to announce the budget cuts and explained that parks would be closed for one day a week. This triggered an uproar of protest. Yet the reality was that Sansom was trapped in the canyon. He was hobbling around on homemade crutches and nowhere near able to hike seven or eight hours up to the rim of the canyon. He couldn't ask other group members to abort the trip and take him to the next raft exit. Continuing the trip also was daunting, with nothing but plastic trash bags wrapped around his leg to keep his foot dry and some antibiotics from the nursing station.

The only option was to get back in the raft. As he did, he was thinking he'd blown the best job he'd ever had—berating himself for going on vacation during critical budget preparations. That night, he lay sleepless on a sandbar, staring up at the infinite number of stars that are visible in remote places. The next morning, the group climbed into their inflatable rafts and soon could hear the roar of rapids that dropped thirty-seven feet below and were two hours away. Even before reaching Lava Falls rapids, the waves were twenty feet high.

Lava Falls is one of the hardest whitewater rapids along the Colorado River, rated as a class ten. At about noon, the group careened over the falls and Sansom was thrown out of his raft, which flipped. Wearing a helmet and a lifejacket, Sansom tumbled downstream, trying to catch a breath before slamming into the next wave. It was one of the few times in Sansom's life that he was thinking about who he hadn't said goodbye to. Eventually, after about a mile of tumbling, he was rescued by another raft.

He lay on the bottom of the raft and moaned, agonizing about what had happened to his buddy, the kindergarten teacher. It was hours before he found out the teacher was fine and the whole group was reunited. That night, he pondered the life-and-death events on the river and the political firestorm back in Austin. After much soul searching, he concluded, "What are they going to do, sue me?" He didn't worry about Parks and Wildlife again until he got home. It was an inflection point where Sansom realized that he had to ground himself in his own being if he ever were to find a more permanent sense of serenity.

The political fallout was still raining down when he returned to Austin, fueled by Bullock's incendiary remark: "He might as well just keep paddling." Yet Sansom kept his job and survived to fight another day—relying on resilience, wits, and allies. His commissioners stood behind him.

Before the next legislative session in 1993, Sansom felt that he was learning how to work with Bullock. He made his pilgrimage to the wood-paneled office in an effort to ingratiate himself with the volatile lieutenant governor and started the meeting with a progress report on one of Bullock's pet projects—the state historic site at Washington-on-the-Brazos. It was badly in need of repair, and Sansom worked with Bullock's brother Tom, an architect from Brenham, to help with restoration. The project was fast-tracked, and the lieutenant governor seemed pleased.

Sansom moved on to his legislative agenda, none of which was especially controversial, and Bullock leapt to his feet. "You are the most uncooperative SOB in state government," he yelled. "You get out of my office and never come back."

Sansom immediately went and bought a pack of cigarettes after being thrown out of Bullock's office for reasons known only to the lieutenant governor. Sansom hadn't smoked in twenty years and was terrified of starting again. To avoid that, he went to a doctor to get a Nicorette prescription and after the nurse took his blood pressure, she left the room. The doctor came in and told him he wasn't leaving because his blood pressure was so high that a stroke was feared. This was two days after the Bullock meeting.

Shortly thereafter, Sansom had occasion to speak with Bass, the commission chairman emeritus. "You're finished," Bass said. "The lieutenant governor can't work with you." Bass was the Fort Worth billionaire who had supported Sansom's idea for the Parks and Wildlife Foundation. Having inherited millions from his uncle Sid Richardson, Bass was a flamboyant oil wildcatter and a major philanthropist across the state, donating to causes ranging from the performing arts to fisheries research.

The following weekend, Sansom was at home with his daughter, April, who was visiting her parents while a student at Texas A&M University. The phone rang and April answered it. The screaming at the other end forced her to hold the phone away from her ear while she said, "Daddy, it's Governor Bullock."

Sansom took the phone and heard Bullock shout, "You sicced Old Man Bass on me!" The bombastic lieutenant governor was furious and had raged at TPWD commissioners as well. By this time, it was becoming apparent to many that Bullock was given to arbitrary and excessive outbursts. The realization saved Sansom for the time being.

While Sansom was lurching from battle to battle with Bullock, he scored a victory at the legislature. He waged a coordinated campaign with two high-profile lawmakers and the TPWD Commission chair to secure more money for the agency through a different kind of tax. The campaign capitalized on a legislative committee recommendation to replace the dwindling revenue from a cigarette tax with more generous revenue from a sporting goods tax. The committee noted that sporting goods better aligned with the mission of Parks and Wildlife than cigarettes. State Senator John T. Montford and Representative Rene Oliveira sponsored the bill to change the revenue source and were strongly supported by TPWD Commission Chairman Nacho Garza, who was close to Oliveira. Sansom was active in the lobbying effort.

The next Bullock blowup came shortly thereafter when the lieutenant governor went to the funeral of former Texas Governor John Connally in June 1993. Connally was laid to rest in the Texas State Cemetery in Austin, which was established in 1851 and located a few blocks east of the capitol amid lovely rolling hills and big oak trees. Bullock became furious about the neglected state of the cemetery and vowed to put things right. He called an interagency meeting of department heads and railed at them for failing to care for the cemetery, even though none of them were responsible for maintaining it.

"He's screaming, 'You people are the worst group of sorry people,'" Sansom recounted to Dan K. Utley for the Texas State University Oral History Project. "'You let this Texas treasure go to pot, and the people of Texas will never forgive you.'" The next day prison convicts were tidying the grounds and bulldozers were cleaning up, yet chaos reigned.

Bullock still wasn't happy. He called the agency heads in for a second meeting and was even more incensed. Sansom was persona non grata around the capitol, following his ouster from Bullock's office, and not allowed to go to the senate except for the finance committee hearings. The lieutenant governor asked the assembled group rhetorically "What would John Connally have done?"

Sansom figured he had little to lose and spoke up. "I think we need a plan," he announced. He could see the horror in his colleagues' faces, signaling their fear of what Bullock might do. "Oh, shit, what is going to happen?" was writ large, Sansom recalled.

Bullock turned to Sansom and brightened up, saying "That's right—do it." So Sansom hired a national expert in cemetery restoration and brought him down to Texas. He had a long ponytail, and Sansom was unsure how that would sit with Bullock. Nevertheless, Sansom figured he was on a roll and decided to risk it. He went ahead and introduced the restoration expert to Bullock, hoping for the best. The expert proceeded to explain how the project would work, and Bullock loved it.

The cemetery restoration was a major turning point in Sansom's relationship with Bullock. Sansom headed up the effort, bringing in Parks and Wildlife's top landscape architect and the Texas Historical Commission. Bullock was thrilled and after that, Sansom could do no wrong.

Bullock rounded up the money for the restoration project, somewhere between $4.5 million and $6 million. "I wish every day that Bullock was still here because I believe in my heart of hearts that as bizarre a character as he was, he had the best interests of the state at heart," Sansom said. "He loved the cultural and natural history of the state, and he would break people's arms to get stuff done." Sansom and Bullock remained lifelong friends for two reasons—Sansom's resilience and his respect for Bullock's political genius.

The choice of Sansom to lead Parks and Wildlife had emboldened agents of change within the agency, particularly those involved with endangered species. Biologists and others viewed Sansom's prior work in acquiring land for protection by the TPWD and The Nature Conservancy as evidence of his priority on environmental conservation.

Some of the agency's top biologists pushed for Parks and Wildlife to get more involved in protecting endangered species by taking over a research program created by The Nature Conservancy and the General Land Office of Texas. The Texas Natural Heritage Program was founded by the two partners with informal support from TPWD for the purpose of studying and preserving endangered species across the state. The program had only a two-year life span and needed a sponsor to continue. Several top TPWD biologists felt the agency should take over the Natural

Heritage Program, which was opposed by private property activists who feared that its database would be used by federal officials to identify endangered species on their lands. Critics accused Sansom of not doing enough to save the program, which was dissolved in 1995.

"The reason I did not continue the Natural Heritage Program is that I caught an employee trespassing on private land to identify endangered species for notation in the program, and I knew that was a surefire way to ruin our careful cultivation of private owners to become better land stewards," Sansom explained.

The employee, who was a biologist, vehemently denied the accusation, according to Wendy Gordon, who has a doctorate in botany and oversaw the program for a while. In fact, the allegations could have come from landowners who were unhappy about TPWD land acquisitions, she said. "The accusations may have been a convenient cover for shuttering the program," she added.

Termination of the program triggered the departure of three scientists from TPWD, which was a "horrible loss," according to Scott Royder of the Lone Star chapter of the Sierra Club. At the time, he said, "Endangered species are now going extinct four hundred times faster than at any other time in recorded history, and yet, Parks and Wildlife is dismantling the very tools we need to address that problem. Now we have no state agency involved in protecting our natural resources."

The end of the program came at a sensitive time. Parks and Wildlife was playing a leading role in the study of the Barton Springs salamander and the Jollyville Plateau salamander for potential listing as endangered species. While the City of Austin paid for the study, Parks and Wildlife provided the personnel to compile the report. The scientists' work was supposed to take a neutral position on listing, though it recommended "establishment of long-term conservation plans and policies to ensure that no intolerable deterioration of water quality occurs in the Barton Creek and Bull Creek watersheds." Another recommendation was to "restrict development in critical areas in the watersheds until data clearly show that continued development will have no further negative effect on water quality."

While Sansom favored environmental conservation, he knew the government would never be able to protect all the resources in need. Landowners would have to be converted to the cause if any protection of endangered species or habitats were to occur.

The idea was challenging in a state where most private landowners at the time viewed regulation with suspicion if not downright hostility. A virulent strain of private property advocates formed the Take Back Texas movement in the mid-1990s, and the zeitgeist helped George W. Bush win his gubernatorial race against Ann Richards in 1995, in Sansom's opinion.

"Andy wisely put the emphasis not on the government, but on private Texans saying we want to make this happen," noted Karl Rove, political consultant and friend of Sansom's. "What are we going to do to protect our great heritage? I think that's a powerful sentiment in Texas, and Andy tapped into it first at The Nature Conservancy and then at Parks and Wildlife."

Sansom's focus on individual responsibility enjoyed support from his Parks and Wildlife commissioners, several of whom became personal friends. One was Ygnacio (Nacho) Garza, who was appointed in 1991 by Governor Ann Richards and became the first Hispanic to chair a board that had been widely viewed as a white men's club for outdoor sports. Shortly after taking over as chair, he was invited to speak at a wildlife conference in Kingsville. "Nacho was a great speaker, and when the program was over, I told him how great I thought his speech was," Sansom recollected. "He said, 'Andy, they're just glad I speak English.'"

Another supportive commissioner was Nolan Ryan, the baseball Hall of Famer who grew up in Alvin, Texas, and was appointed by Governor George Bush in 1995. A few years into Ryan's six-year term, he and Sansom traveled together to investigate some wildlife issues in South Texas. On the way back to Austin, they stopped for lunch at a restaurant in Refugio, where Ryan was born. "When we finished lunch, we were standing at the register to pay our bill and the cashier reached into the cash drawer and pulled out a very collectible Nolan Ryan baseball card," Sansom recounted. "She said, 'You know who you look like?' and Nolan responded, 'Yeah, people tell me that.' She then said, 'You know he was born here?' and Nolan said, 'I heard that.'

"She said, 'My son pitches in Little League, and he told me if Nolan Ryan ever came in here, I was to get him to sign this card,'" Sansom continued. "Whereupon Nolan replied, 'Now's your chance!' and she exclaimed, 'No! I want Nolan Ryan to sign it.' Nolan said, 'OK Andy, let's go,' and we headed for the door. One of the waitresses cried, 'Janie, it's him,' and the cashier reluctantly allowed him to sign the card."

While Sansom benefited greatly from the political backing of his commissioners, he had to have practical tools to implement his policy objectives. One of his most valuable tools was the conservation easement, which made its way into Texas law in 1983 after the federal Highway Beautification Act provided "scenic easements" as a model. The intent was to protect the land's natural, productive, or cultural features in exchange for tax benefits to the landowner.

Sansom recognized that Texas pride could be the key to conservation easements in the Lone Star State. Landowners typically took great satisfaction in the possession of their property, prompting them to want to keep it in their families, even at the price of foregoing development. The prospect of lower taxes and the voluntary nature of the pacts overcame skepticism about the government entities that were counterparties to the covenants.

"Andy's philosophy is that you have to respect private property rights," said Francell, who worked on conservation easements for TPWD the first time he was hired at the agency in the early 1990s. "If you want to protect it, you buy it," he said, referring to Sansom's push to purchase critical lands. "And when you buy it, you have to have a willing seller on the other side."

Sansom was "pretty solidly Republican" at that time, noted Francell, who was hired by Sansom. Francell supported the notion that landowners should be compensated through lower taxes for giving up their development rights. "You've got to bring them to the table with something," Francell explained. "The idea of paying for conservation easements fit perfectly with his mentality."

By the mid-1990s, Sansom was at the top of his game. He felt confident enough to crack down on sexual harassment at the male-dominated agency through more intensive training and dismissal of employees who flouted policy guidelines. One of those fired for sexual misconduct had a flamboyant personality and was loved by outdoors writers because he was a colorful character with a large handlebar mustache.

He held a high position in the law enforcement division and had an administrative assistant who was a young woman. Her cubicle was just outside his office. Immediately after being hired, she had hung pictures of bluebonnets and cuddly animals around her desk. "She comes in the next day and the pictures are gone and there's a deer head hanging over her desk," Sansom recounted. "She sat there all day, wondering what to do. Finally, she took the deer head down, put it back in his office, and put her pictures back up. The next day the deer head was back. She went into his office and said, 'Sir, I feel like this is your space and mine is out there. I like to have things around me that have meaning for me.' He just laughed her off."

She filed a complaint against him, which turned out to be the tip of the iceberg of sexual harassment. "I fired him," Sansom continued. "His behavior today would be considered egregious, but in those days, it was just routine. Even some of the women had just long accepted that behavior. One of the aspects of change that I found the most painful was doing things that were anathema to people internally. My assistant, who was a female and had been there twenty-five years, came into my office furious with me for firing the popular guy. She didn't forgive me for a long time because that was the way it was."

Sansom had restored agency morale following the political scandals surrounding wildlife stocking for influential landowners and lawmakers. Parks and Wildlife had among the lowest turnover rates in state government because employees liked their jobs.

As a manager, Sansom was liked and respected by employees, and many of his closest friendships were formed during that time. His outgoing nature, signature chortle, warm style, and passion for the job won over many of them. Those who opposed him found that Sansom would listen to both sides of an issue before making a considered decision. He mediated the historical tensions between park staffers and wildlife staffers. When employees were unhappy, they left. The biggest criticism that some employees had was that Sansom's dreams exceeded his ability to implement them.

Those dreams proved to be his undoing.

Chapter 14

Caught in the Cross Winds

When George W. Bush was elected governor of Texas in 1995, he was only the third Republican to hold that position in 126 years. He had several things going for him. Name familiarity was one—his father, George H. W. Bush, had served as US president only six years earlier. Another was a shrewd political strategist, Karl Rove, who created an image of "personal responsibility" and "moral leadership" for Bush to replace the earlier one of a playboy who, in his own words, "chased a lot of pussy and drank a lot of whiskey." Moreover, the conservative tilt of all Texas politics meant that party labels could be swapped out as long as the candidate hewed to the right of center.

Rove believed the electorate wanted welfare reform, tough-on-crime policies, and more local control for schools. In fact, he is often credited with turning Texas red in the 1990s after more than a century of blue dominance. "It was just a matter of being at the right place at the right time," Rove demurred.

A master mixer of campaign strategy and governing policy, Rove wanted to ensure that Sansom got along with the incoming governor following his close relationship with the outgoing one, Richards. During the transition between the two administrations, Sansom and Rove took their annual quail hunting trip to the historic Kenedy Ranch in South Texas, where aristocrats have fraternized over dead animals and luxury libations for decades.

Sansom had learned to hunt birds as a youth, traipsing through coastal marshes with his buddy Corky Palmer, who knew about shotguns and hunting dogs. Palmer's father owned the local sporting goods store in Lake Jackson, and Palmer owned a Labrador retriever, a dog that cemented Sansom's love of canines after the loss of his family's roast-eating boxer.

"Hunting is largely about the relationships with people you love," Sansom explained. "Hunting can be a compelling opportunity for fundraising, and I began to seriously hunt as an adult with prospective donors and legislators. For me, hunting—particularly for birds—is all about being in the company of dogs, experiencing their joy and their skill."

In 1995, Sansom understood the necessity of working well with Bush and saw no obstacles, given that both were lifelong Republicans and personal friends of Rove. Since meeting in Houston in the seventies, the Roves and the Sansoms spent time together as couples, and Rove named his only son, Andrew, after Sansom.

Governor Bush and Sansom broadly aligned on environmental policy as they both embraced the Republican dogma of the day—small government, low taxes, and minimal regulation. While the federal and state governments had a clear role to play in supporting parks, wilderness areas, and wildlife refuges, the private sector ruled in Texas. Property rights trumped most other policy, given that the overwhelming majority of land in the state was in private hands. Conservation programs required landowner acquiescence, which was likelier to be obtained through tax policy and financial incentives than mandates.

Sansom and Bush could draw on the storied history of environmental conservation in the Republican Party, going back to Teddy Roosevelt, who established 150 national forests, created five national parks, and protected more than 230 million acres of public land. Senator Barry Goldwater advocated for environmental conservation in his 1970 book, *The Conscience of a Majority*, arguing, "Our job is to prevent that lush orb known as the Earth . . . from turning into a bleak and barren, dirty brown planet." He continued: "Although I am a great believer in the free, competitive enterprise system and all that it entails, I am an even stronger believer in the right of our people to live in clean and pollution-free environments."

The most influential Republican in environmental legislation was Richard Nixon. In the 1970s, he had created the Environmental Protection Agency and signed into law more major pieces of environmental legislation than any president before or since. By the 1980s, the Republican Party platform referred to government regulation that was "firmly grounded on the best scientific evidence available" and laws "that are enforced evenhandedly and predictably."

But by that time, business and industry had started rebelling against regulations that were deemed damaging to profits and international competitiveness, thereby undermining the GOP's historical support for environmental policy. Conservatives argued that the federal government's power should be limited to taxing, declaring war, regulating foreign commerce, and not much more. Most responsibility should reside with state and local governments and even then, on a small scale. Market forces should be unleashed to solve the nation's ills.

Nevertheless, in 1988, George H. W. Bush pledged to be the "environmental president" and subsequently signed into law the Clean Air Act Amendments of 1990, the most expansive and possibly most expensive piece of environmental legislation in the nation's history. When his son "W" ran for president, the younger Bush warned against ceding too much power to Washington. The negative consequences of slower growth and greater regulation resulting from environmental legislation far outweighed any ecological risks, he argued. Moreover, individual liberties were jeopardized by government intervention. While Bush's laissez-faire approach to the environment might have been less helpful than Sansom occasionally would have liked, the two Republicans shared a similar philosophy on social issues. Bush ran for president as a "compassionate conservative," and Sansom tried to live that credo.

Based on shared principles, the area where Sansom and Bush made the most lasting impact was in water. In 1997, landmark legislation to create a state water plan for Texas was sponsored by state Senator Buster Brown, whose first election campaign was chaired by Sansom. Governor Bush threw his weight behind the legislation, Senate Bill 1, and the legislation passed. The law created a bottom-up process that enabled each of sixteen regions across the state to evaluate its water supplies, needs, and shortfalls and to identify projects to fill the gaps. The regional plans were then rolled up into the state water plan.

To get the backing needed for the legislation, Sansom and Brown worked with Bush's environment and natural resources policy director, John Howard. "Governor Bush's approach was largely hands off," explained Howard. "He did not meddle. He understood what was going on, but he didn't say you have to do this. He saw his role as making sure that he had the best people, and then they could do their thing and get legislative support."

On other water fronts, the issues were more challenging. Endangered species, for example, triggered a political uprising in the mid-1990s. Landowners feared that endangered species would be found on their property and bring federal regulators onto their land. These fears gave rise to the Take Back Texas movement, a grassroots group led by a flamboyant native son named Marshall E. Kuykendall Sr. A descendant of the original three hundred Anglo families allowed by Mexico to settle in Texas under Stephen F. Austin's charter, Kuykendall opposed the federal government's "invasion" of private property.

Despite the property rights activism, Bush and Sansom produced a seminal report called "Taking Care of Texas," which was a master plan for environmental conservation in Texas with a blueprint on how to implement it. A governor-appointed task force, including Samson, produced the plan. "Texas needs a more comprehensive, science-driven strategy for the conservation of its outdoor resources," declared the task force members.

Some of the key recommendations were to:

- Create a statewide program to purchase development rights from landowners.
- Develop a comprehensive system to assess conservation and outdoor recreation needs.
- Acquire assets that meet criteria of statewide significance to meet those needs.
- Ensure adequate quantity and quality of water to protect its land and water ecosystems.
- Expand incentives for habitat management and outdoor recreation on private lands.

The most progressive recommendation was to buy development rights from landowners, as that approach would go beyond conservation easements that only provide tax relief as compensation to property owners. "We didn't know exactly what that program should look like," admitted Howard. "It was very pioneering, and Andy saw the opportunity for Texas to lead the way."

While some saw the report as pioneering, others viewed it as a bridge too far, even those in Governor Bush's party. The task force was "chock full of Republicans," noted parks advocate George Bristol, who tirelessly lobbied for more money for parks. "It wasn't exactly a tree-hugging, radical group of people that made these recommendations."

Still, "it stirred controversy," Bristol continued. When the recommendations reached legislators, "they put the kibosh on it." Sansom was caught in the crosswinds, slammed for daring to protect Texas springs, help landowners manage habitat, and encourage nature tourism.

The antiregulatory sentiment that pervaded the capitol also fueled other discontent, such as distrust of federal bureaucracy in general, skepticism about science, and denial of climate change. These forces gathered pace when Rick Perry, then lieutenant governor of Texas, ascended to the governorship in December 2000 after Bush won the presidential election.

The 2001 legislative session was brutal for Sansom, even by the standards of a legislature that was often critical of Parks and Wildlife. House Appropriations Committee Chair Rob Junell lambasted Sansom during one particularly painful hearing, leaving Sansom to stagger out after hours of grilling.

Despite the legislative inquisitions, Sansom's name was being floated at the federal level for a high-profile position. He was put forward as a candidate to head the National Park Service and seemed like a strong contender, given that he had served under President Bush for five years when he was governor. In addition, Sansom was close to Karl Rove, by now a senior advisor and deputy chief of staff for President Bush.

Sansom's candidacy was vigorously supported by Bristol, who had gotten a phone call from fellow parks advocate and philanthropist David Rockefeller Jr., inquiring about Sansom. Bristol and Rockefeller discussed other candidates as well and agreed to support whomever Bush nominated. Behind the scenes, opposition to Sansom made its way to Bush's people. State Representative Junell, who had clashed with Sansom over funding, may have weighed in against Sansom. Likewise, Texas House Speaker Pete Laney, who was close to Junell, may have given thumbs down.

In the end, a dark horse candidate from Florida was chosen, following lobbying by Jeb Bush, then governor of Florida and the president's brother. Sansom accepted the outcome with grace and bore no grudges, in his inimical way.

Nevertheless, Perry was settling into the big, white governor's mansion across from the capitol and setting his political agenda. It was clearly to the right of Bush's and embraced issues that were beyond classic Republican ones. Perry viewed climate change as "unsettled science" and opposed the regulation of greenhouse gases as pollutants, suing the administration of then President Barak Obama to overturn Environmental Protection Agency regulations. He advocated for the teaching of creationism in public schools, and at one point, proposed putting the Texas Parks and Wildlife Department under the Texas Department of Agriculture. The idea was viewed as a bit dotty.

Sansom was troubled by some of Perry's positions, such as demonizing the federal government, denying that human activities contribute to climate change, and seeking to undo protections for the environment. "It all changed with Perry," Sansom told the *Texas Observer* in 2018.

Moreover, Perry signaled early on that he wanted to widen the governor's powers. He aimed to put his own people in top jobs at state agencies—a departure from past practice where agency commissions typically chose their own executive directors. Texas Parks and Wildlife was a case in point. In 2001, Perry appointed Katharine Armstrong Isdal, a Texas aristocrat of the first order, to chair the TPWD commission. She was the heiress to two ranching fortunes of Texas—the Armstrong and King dynasties, built on land, cattle, banking, and oil—and her mother was an ambassador to the Court of St. James's.

The King Ranch, founded by Richard King in 1857, is 825,000 acres, making it the largest ranch in the state and bigger than Rhode Island. The South Texas spread served as the model for Edna Ferber's novel of Texas aristocracy, *Giant*.

The nearby Armstrong Ranch is more than 50,000 acres and was a quail hunting playground for the rich and famous from Texas and beyond. Guides find the birds, and trained dogs flush, point, and fetch. After the quail killing, cocktails and barbecue abound.

As the new commission chair, Armstrong invited Sansom to breakfast at the Four Seasons Hotel in Austin. Former chair Lee Bass also was present. Without much ado, Armstrong announced that the governor wanted Sansom replaced with someone of his own choosing.

Sansom was blindsided, though he remained outwardly calm. He quickly collected his thoughts and decided that he would agree to leave, though not until the end of the year. It was August. It was a clean and quick death by dagger. After Armstrong left, Bass gave Sansom a hug. The commissioner indicated that he still felt very close to Sansom, who went home and talked to Nona.

"Andy came home and said, 'Rick Perry is trying to fire me!'" Nona recounted. "It was the first use of the F word in this situation. 'But he can't do that! It is the commissioners who hire and fire the executive director, not the governor. If I called a meeting of all nine commissioners on this issue, six of them would vote to keep me on the job.' I asked, 'Do you really want to work for Rick Perry and Katharine Armstrong?' The answer was, 'No!'"

A self-aware person, Sansom knew that he'd made enemies over the nearly twelve years that he was executive director. Every decision he'd made caused some people to be happy and others to be angry. "It's the mad ones that accumulate," Sansom said pensively. "I was getting pretty beat up, particularly in the legislature. Whereas in the beginning, I was the fair-haired boy."

While governors and other elected officials often want to clean house and install their own people when entering office, Perry delegated the dirty work to Armstrong. "She took it upon herself to purge all Andy's friends and vestiges of his time with the agency," Bristol wrote in his book, *On Politics and Parks*. Bristol continued, saying that Armstrong's "purge extended beyond the agency to other organizations, such as members of the staff and the board of the Texas Parks and Wildlife Foundation. At some point, even photos of Andy, formal and informal, disappeared from the walls and halls of TPWD."

Ironically, it was after Sansom was dismissed that one of his longest-running labors bore fruit. The funding he had sought for deferred parks maintenance was approved by voters in 2001, making it the biggest bond issue in the history of Parks and Wildlife. The measure authorized the issuance of up to $850 million in general obligation bonds for construction and capital repairs. The ballot box victory came a month before his last day at the agency.

Over the eleven-plus years Sansom had been at TPWD, he doubled the amount of parkland owned by the agency, put parks infrastructure on a more equal footing with wildlife, and moved agency culture beyond a hunting and fishing club for white men.

He was generally seen as kind, considerate, and thoughtful—he asked about family members, remembered names of staffers, mentored employees, and listened carefully. Sansom kept open lines of communication with direct reports and was willing to "give folks the ability to take charge of their divisions" without second-guessing them, recalled Dee Halliburton, who worked for TPWD for more than thirty-five years. She was an executive assistant to four executive directors, including Sansom, and called him a "great man."

Sansom also hired staff members to strengthen areas that he felt were weak. Energy and pragmatism enabled him to push through entrepreneurial ideas for making TPWD more efficient. "He was seen as a charismatic leader, a visionary," said Francell, who ran the land acquisition unit. "He had great ideas and was definitely intelligent."

Sansom was adept at balancing competing interests and finding common ground. He was not a long, tall Texan full of bluff and bluster, but rather a medium-sized one with a winsome smile and penchant for pragmatism. His mild manner accentuated his reasonableness, and he had a hard time saying no. Sansom lived intentionally and used humor to disarm opponents.

After nearly twelve years, though, Sansom was beleaguered. When he started, he was the fair-haired boy with exciting ideas. By the end, he was staggering out of legislative hearings that felt like interrogations. He was being criticized for acquiring too much land for the state. Big Bend Ranch State Park was still a sore subject for many lawmakers, some twelve years after the deal was done.

Sansom was at TPWD for more than eleven years, only two months shy of the longest serving executive directory in the agency's history. The average tenure when he started was eighteen months—and he had expected to stay for no more than five years.

"Even if Katharine Armstrong had not come along, I was weary," Sansom said wistfully.

Chapter 15

Aquamaids and Flying Pigs

For more than ten thousand years, people have been living around the San Marcos Springs because of the abundant water and food. In prehistoric times, the artesian springs provided ample water for drinking and cooking, while reeds and grasses along the waterways attracted wildlife that fed the hunter-gatherer people.

Some two hundred springs bubble up from the Edwards Aquifer and course through fissures and small openings in the bottom of Spring Lake, which then spills into the headwaters of the San Marcos River. According to early accounts, the artesian springs once spurted up several feet above the surface. In the seventeenth century, a Spanish military expedition sent to destroy French settlements in Texas discovered the river between Austin and San Antonio and christened it the San Marcos in honor of Saint Mark, whose feast day had just passed. Thereafter, the enticing river attracted an array of settlers—German peasants, Anglo herdsmen, and Texas entrepreneurs. In 1849, an enterprising former vice president of the Republic of Texas, Edward Burleson, built a dam downstream from the springs to power a gristmill and sawmill. The reservoir behind the dam was named Spring Lake.

Nearly eighty years later, in 1926, San Marcos businessman A. B. Rogers decided to commercially develop the lake. He bought 125 acres surrounding the crystal-clear waters and built a luxury hotel in art deco style overlooking the springs and lush vegetation, hosting swimsuit beauty contests and swanky parties on the rooftop. In 1946, he added glass-bottom boats so guests could see fish, turtles, and grasses in their native underwater environment.

By 1950, Rogers's son took over the operation and expanded it into an amusement park called Aquarena Springs. Paul Rogers modeled his park after Florida's Weeki Wachee Springs, which had pioneered the use of hidden air hoses to allow entertainers to breathe under water. Rogers dredged part of the lake to build a 125-seat viewing gallery that was billed as the "World's Only Submarine Theater." A mechanical contraption that made the cover of *Popular Mechanics* magazine in 1952, it submerged spectators twelve feet underwater to watch aqueous performances. Women dressed as "Aquamaids" had picnics underwater and frolicked with Glurpo, the swimming clown.

One unique part of the show was Ralph, the flying pig. The porcine entertainer would run down a ramp on a humanmade volcano and "swine dive" into the lake, then swim, as the theater slowly submerged. A clever marketer, Rogers hosted the first underwater wedding, which was covered in *Life* magazine. A Swiss-designed gondola gave guests an unprecedented view of the springs from high above.

By the 1970s, however, the park was struggling as attendance fell victim to high gasoline prices and competition from newer Texas tourist attractions such as Six Flags over Texas and SeaWorld San Antonio. 1985, the Paul J. Rogers Trust sold the property to the Baugh/Moore Investment Company, which continued to run Aquarena Springs for a while.

At the same time, concern was growing about the environmental sensitivity of the springs, the second largest in Texas after Comal Springs. The San Marcos Springs, the lake, and the river ecosystems are environmentally significant, with the upper reach of the river among the most biologically diverse aquatic systems in the southwestern part of the United States. The San Marcos River Foundation was created in 1985 to protect the river from poorly treated wastewater that had been dumped into it for years and amid plans for largescale water pumping from it.

A couple of years after Baugh/Moore bought the park, the investment company was eager to sell the property, and one of the owners contacted Sansom, who was the landman for Texas Parks and Wildlife at the time. Sansom discussed the possible acquisition with his commission chairman, Chuck Nash of San Marcos, and they met with J. Lloyd Moore, co-owner of the investment company, at the old restaurant at Aquarena Springs. Ultimately, however, TPWD decided that Aquarena Springs did not fit the agency's definition of a state park.

Watching carefully from the sidelines was Aquarena's neighbor, Southwest Texas State University (now Texas State University) and its president, Dr. Jerry Supple. A chemist by training, Supple understood that the San Marcos Springs were environmentally significant, one of the biggest artesian springs in the United States. In 1980, the San Marcos Springs were declared critical habitat by the US Fish and Wildlife Service when the San Marcos salamander was designated a threatened species. Thereafter, another six species of plants and animals in the vicinity were designated as endangered.

Supple feared what might happen if the school didn't step in, and he recognized that owning such an environmental gem would be unique for any university, especially in Texas, where he wanted his institution to be more competitive. While he could appreciate the environmental value of the springs, his board of regents was less convinced. The regents were skeptical of buying property that couldn't house a classroom or dormitory, and local civic leaders worried about the loss of tax revenue alongside city merchants who feared a drop in income if the amusement park went away.

It took until 1994 for Supple to overcome resistance and persuade the university to acquire the property from the Baugh family, which had assumed ownership by then. The university paid about $8 million for the park and then invited its own Texas Rivers Center, the City of San Marcos, the US Fish and Wildlife, and the US Army Corps of Engineers to jointly develop a plan for the park's future. Aquarena Springs continued as a theme park for a while, though business tapered off amid mounting problems with the infrastructure, including fears that the dam holding back Spring Lake would crumble. The park had been built in a flood-prone area, and the structures contained asbestos and lead paint, which were routine materials seventy years earlier but deemed toxic in the late twentieth century. Costly remediation was required.

During this time, Sansom—who was running Parks and Wildlife— offered to partner with the university to convert the amusement park into an education center for the public, modeled after TPWD's Sea Center in Lake Jackson and its Texas Freshwater Fisheries Center in Athens. His offer was accepted and the park was closed in 1996.

In 1999, discussions began with the Corps of Engineers to restore Spring Lake to its more natural state, similar to before the amusement park was built. The project would end up taking nearly twelve years. Supple was looking for money and leadership to use the property for expanding the university's water programs with more robust education and research. He attended a House Appropriations Committee meeting, hoping to find some public funds, and watched Sansom undergo a particularly brutal line of questioning about Parks and Wildlife discretionary initiatives, budget cutting, and fee increases.

After the grueling hearing, Sansom trudged out, and Supple followed him. "If you ever get tired of this, you can come down to Southwest Texas," Supple told Sansom. Sansom didn't take it seriously at the time, though he had second thoughts six months later as the political winds shifted against him. Sansom left Parks and Wildlife at the end of 2001, and by early 2002 he had moved to Southwest Texas State University.

It was his second foray into academia after being at the University of Houston in the 1970s, which had left a bad taste in his mouth because of internal squabbles. "Politics in universities are the worst because the stakes are the smallest," Sansom said, citing a cliché in the academy. His arrival in San Marcos came at an auspicious time.

The university had just won a grant to create an International Institute for Sustainable Water Resources, based on a proposal by world-renowned limnologist Walter Rast. He wanted to elevate water research as a valued discipline within the university, and Supple asked the Meadows Foundation and Houston Endowment to fund his plan. Still, Rast didn't want to run the institute because of an aversion to politics, according to a former graduate student at the university who was close to him. The institute was the first of its kind at the university, which historically had specialized in the education of public school teachers. The Sustainable Water Resources Institute was designed to teach, conduct research, and offer service activities. Sansom's job was to implement Rast's vision, raising the necessary funds from both nonprofits combined.

It was a defining moment in Sansom's career, marking a clear pivot to water as the central focus of his work. For the rest of Sansom's professional life, water would be a guiding star.

Earlier, while still at Parks and Wildlife, Sansom had agreed to the $3.5 million conversion of the hotel and surrounding buildings into headquarters for the Texas Rivers Center, including academic and research facilities. The project was funded 35 percent by the university and 65 percent by the US Army Corps of Engineers, with Parks and Wildlife participating on a barter basis. The agency contributed to the restoration, and in return received free space for staff members for more than a decade. The agreement also called for the university to donate rights to forty thousand acre-feet of water annually to the Texas Water Trust, which holds rights for environmental purposes.

At the same time, Sansom needed to establish the university's bone fides in water research and policy. Texas State University had been viewed by some as a party school with ample opportunity for carousing on the river and less rigorous academic standards. About a quarter of students chose the university for the springs and river, according to Sansom. "What's wrong with that?" he asked rhetorically, since rivers weave together recreation and appreciation of nature.

Sansom believed the springs and lake should be a magnet for students seeking a unique environmental education through an outdoor laboratory right out the front door. Students should be able to sketch a largemouth bass or dissect its liver while overlooking the lake from a boardwalk, for example.

The ecosystem had the potential to set Texas State University apart from the state's tier-one universities, Sansom said at the time, referring to those that aim to attract at least $100 million in research grants. "Having the headwaters of the San Marcos River on campus and having strong aquatic resource programs in the biology and geography departments make water a major part of the school's culture," he said. Equally important were primary and secondary school students, as Sansom had long supported efforts to get kids outdoors. He envisaged schools across the state coming to San Marcos for field trips.

For Sansom, the job was an ideal platform to widen his influence over water policy in Texas. He testified at the legislature, spoke at industry conferences, served on task forces and commissions, earned a doctorate degree, and set himself up as an industry expert with the media. After decades of being in the spotlight, he was good at navigating the media landscape and cultivating relationships with journalists. Most importantly, Sansom was comfortable moving in and out of various spheres of power.

To establish the institute's credentials within the university and beyond as a player in the policy world, Sansom was keenly aware of the need to build a reputable staff. His approach was to mold the institute's work around the strengths of his experts, and a natural place to look for talent was his previous job. At Texas Parks and Wildlife, Sansom had hired a bright young intern who held master's degrees in both environmental science and public affairs. Her name was Emily Warren Armitano. "She absolutely flourished," Sansom said. "By the time I left, Emily had written the statewide land and water conservation plan. She was recognized as one of the real brains at Parks and Wildlife."

In 2003, when Sansom reached out to her, Armitano was policy and regulatory coordinator at TPWD. He persuaded her to come to the fledging Institute for Sustainable Water Resources to be assistant director. Within months she was functioning as the interim director, filling in for Sansom when he was recovering from his car accident. It was a formative time for the institute. "She, probably more than me, is responsible for putting together the building blocks of what is today the Meadows Center," Sansom said. "She hired most of the people that are there today."

The Texas Parks and Wildlife Department was a rich vein for Sansom to mine, particularly the water resources program that he'd created. The leader of that program was a bright and energetic hydrologist named Cindy Loeffler, who had spent years studying how the health of rivers and estuaries impacts fish and wildlife. On behalf of TPWD, she coordinated closely with the Texas Commission on Environmental Quality and the Texas Water Development Board to develop state methodologies for determining what constitutes healthy waterways.

Loeffler told Sansom that he should hire her TPWD colleague Warren Pulich to come to San Marcos. Pulich was a coastal wetlands biologist who had done groundbreaking research in water that courses down rivers and streams and into bays and estuaries, known as environmental flows. It was an area that Sansom wanted to develop at the institute. Sansom called Pulich and said, "The job is wide open," according to Pulich. "He said, 'Start working on things that you're familiar with, in your area of specialization, and we'll see what you can develop.'"

Sansom had told the university that environmental flows would be a top priority of the institute, though Pulich sometimes struggled to see a plan to make that happen. It took a while to get programs underway. "It was an evolving thing. [Andy] is kind of unfocused; he wants to do everything." Pulich felt he was being pulled in several directions at once. "Andy was telling me, 'Why don't you do that? That would be good,'"

Pulich explained, wondering to himself what wouldn't get done if the new task was started. "I just found it hard to finish some of these jobs before I was moving on to something else. He wanted to do everything, and I sometimes found it difficult to understand the priorities for me." The institute could have contributed more to the environmental flows process established by the state legislature in 2007 to develop recommendations and standards for water flowing in all Texas river basins and estuaries, in Pulich's opinion. The process was driven by state-appointed stakeholder committees as well as science panels, where technical data was considered.

Sansom saw it differently, noting that in 2003 he was appointed by Lieutenant Governor David Dewhurst to the Study Commission on Environmental Flows. Its charge was to "determine whether future legislatures should authorize the issuance of environmental flow permits." In addition, Sansom explained that he hired the first endowed professor at the River Systems Institute, Dr. Thomas Hardy, a nationally recognized authority on environmental flows.

Still, Pulich thought Sansom may not have made the most of his inside track to state lawmakers and other agency leaders. "He had the door open to state agency people and lawmakers, they would listen to him. He probably did, and may have felt they were sandbagging."

Sansom's first major project for the institute was landing a contract to lead a science advisory committee for the LCRA-SAWS Water Project, which was a proposal to export water from the lower Colorado River to the city of San Antonio. The San Antonio Water System aimed to buy the water from the Lower Colorado River Authority if technical studies proved the project feasible. The institute's job was to vet the studies for technical integrity, based on regular reviews of project scopes and deliverables, and provide feedback.

In May 2005, Sansom's plate was more than full with staffing, funding, and administration duties to get the institute up and running. In the blink of an eye, though, his life changed irrevocably. An oncoming car swerved in front of him, causing a devastating crash that sliced off his ear, lacerated his face, injured his leg, and left him with months of convalescence. For Sansom, this brush with death lessened the mystery surrounding it.

"It brings home to you that you have a limited amount of time, and you better make good use of it in terms of your relationships with the people you love, in terms of work that you want to accomplish, et cetera," Sansom acknowledged. "It also had a very significant spiritual impact on me." Shortly before the accident, Sansom had decided that he wanted his funeral or memorial service to be held in a church and nowhere else because that's how he wanted to be remembered. The revelation had come after he attended a memorial service at a funeral home and realized that the secular setting lacked the religious and spiritual dimensions that he had cherished earlier in his life.

"It was largely because of my accident that when I got well and was able to get around again, we began returning to the Presbyterian church," Sansom explained. "I had just let it slip away." Church membership rekindled his love for choral singing, his desire to help the less fortunate, and his enjoyment of a faith community. "It was a kind of returning home," he said.

Despite the enormous challenges of recovery, Sansom was able to keep a steady hand on the tiller, with Armitano's help. She served as his envoy, attending conference calls in his stead, relaying information back to him, and presenting his views as needed. After Sansom recovered and returned to work, he continued to delegate freely to Armitano, and he admitted that she carried the water while he gladhanded donors.

A particular priority of Sansom's was the institute's Texas Stream Team, an existing program that trained "citizen scientists" to collect water samples from their local lakes and streams to monitor water quality. The data provided "early warning signs" of potentially polluted water that could be relayed to regulatory authorities and analyzed for best management practices. The Stream Team became a hallmark of the institute and had trained eleven thousand community scientists by 2020. In addition, the Stream Team established Texas Status University's credentials with the Texas Commission on Environmental Quality, which had to approve the university's implementation of the program.

Not long after Armitano's arrival, she and Sansom started discussing whether the institute should focus on rivers rather than sustainable water resources. Rivers are more tangible—through the history, commerce, and culture of communities—and visible in a way that groundwater isn't, because it's invisible until reaching the surface.

In 2005, the institute's staff of half a dozen plus some geography department members held a retreat and decided to change the institute's name to River Systems Institute. More importantly, the institute was given responsibility for managing San Marcos Springs and moved its headquarters from the tiny house in San Marcos to the restored Aquarena Springs Inn, now renamed Spring Lake Hall. "That was huge," Sansom noted. "We had gained sufficient credibility and respect inside the institution that they gave us responsibility for the springs, which are one of the most unique natural sites in the United States."

The name change didn't much affect the institute's work, though it improved the institute's marketability. Sansom, always sensitive to branding, realized the name "International Institute for Sustainable Water Resources" was too long to be remembered and misleading about the geographic scope of its operation. His goal of raising the institute's public profile required a more arresting image, such as rivers and streams. The institute hosted water conferences regularly for academics, regulators, engineers, and others to showcase the new facilities, highlight the institute's work, and increase name familiarity.

One of Sansom's strengths was recognizing his weaknesses and compensating for them by surrounding himself with experts in those areas. "He was always working to keep people around him who were ranchers, industrialists, those who were outside his area of expertise," recalled a former grad student. Sansom was a good leader. "He had daily conversations with everyone," she explained. "He cultivated staff relations."

While Sansom juggled disjointed projects, sporadic funding, and rotating staff, he found the most personal reward in the center's educational program for military veterans. Those who came back from Afghanistan and Iraq with amputated limbs were suffering not only with physical injury, but also emotional wounds. "They felt like their lives were over" when they arrived at the River Systems Institute, Sansom recalled.

The center trained the amputees to scuba dive so they could collect endangered species and repair scientific instruments at the bottom of Spring Lake. "It gave them a purpose," Sansom said, "which the people from the Brooke Army Medical Center in San Antonio say is the most important thing. Does that rank high on the university's list of priorities? No, but it makes going to work every day worthwhile."

Chapter 16

Meadows Foundation Changes the Game

In 1928, a year before the Great Depression began, an attorney and accountant by the name of Algur H. Meadows and a couple of friends founded a loan company in Shreveport, Louisiana. As the economy collapsed over the ensuing years, the resourceful founders devised a clever way to finance oil drilling deals by buying future "production payments" that entitled them to the proceeds of some of the crude produced by the property. In exchange, they provided immediate funds to the seller of the property, relieving the buyer of that obligation to pay upfront.

Meadows is credited with inventing the financing scheme, which was called an "ABC Transaction" because three parties were involved: buyer, seller, and financier. The scheme led to explosive growth for the loan company, which became an oil company after a merger with the General American Oil Company of Texas in 1936. The new company moved its headquarters to Dallas the following year and subsequently expanded and shrank operations before sealing its success with a foothold in the East Texas oilfield. Using "ABC Transaction" financing, the General American Oil Company of Texas proved highly profitable because it could buy oil and gas in the ground and avoid the costs of exploration. Rapid expansion followed, and by 1948 the company owned four hundred wells producing seven thousand barrels of oil daily. Meadows was a wealthy man by the age of forty-six.

At that relatively young age, Meadows and his wife, Virginia, were thinking beyond their years. In 1948, they established the Meadows Foundation to do "good in Texas," and for the next forty or so years it funded projects in the arts, culture, and education to improve the quality of life in communities. *Life* magazine called Meadows "a virtuous millionaire."

During the 1980s, the foundation made several grants to the Texas chapter of The Nature Conservancy, though culture and education remained core areas for funding. While Sansom was running the Texas chapter at the time, he didn't personally meet any Meadows staff. By the early 1990s, the foundation started diversifying into heritage tourism and ecotourism with its philanthropy, starting with the Los Caminos del Rio Heritage Project—a corridor along the Rio Grande with historical, cultural, and natural features. The goal of the grant funding was as much to promote economic growth as to protect nature, hence a birding project was designed to draw visitors rather than directly benefit the avifauna. Still, the Rio Grande Valley project paved the way for a greater emphasis on the environment.

Around this time, the foundation's chief strategist, Bruce Esterline, met Sansom at a Los Caminos event in Roma and was immediately "captivated by Andy, as most people are." Sansom's work with The Nature Conservancy and Texas Parks and Wildlife, his vast network of contacts, and his warm personality made him a valuable advisor to the Meadows Foundation, albeit an informal one.

By the late 1990s, the foundation's third generation of family members made it clear that they wanted to focus more on the environment, telling the older generation, "As a foundation serving Texas, we need to be talking about the environment." Foundation staff were charged with coming up with a blueprint to make that happen, resulting in an initial plan in 2001. By the early aughts, the foundation became one of the biggest philanthropic funders of environmental projects in the state. At the same time, Sansom's advisory role was becoming even more strategic, including the review of internal ideas with staff.

In 2002, Sansom's move to Southwest Texas State University (which changed its name the following year to Texas State University) and the launching of the International Institute for Sustainable Water Resources provided an ideal opportunity for the Meadows Foundation to make a high-profile donation in service of the environment.

Consequently, Meadows and the Houston Endowment teamed up, as they often did. Each provided financial support to get the institute up and running and advance the university's efforts in aquatic resource management. Four years later, Meadows gave another $610,000 to continue the institute's work under its new name, the River Systems Institute.

By the end of the aughts, funding for the River Systems Institute was running low and Esterline had indicated to Sansom that one more grant would be forthcoming. Sansom met with Esterline and Environmental Program Manager Mike McCoy and made his pitch for a final tranche of money. "Esterline told me, 'You're thinking too small,'" Sansom recollected. He hadn't dreamed that Meadows might elevate the River Systems Institute to the level of its two high-profile philanthropic entities that carried out its mission—the Meadows School of the Arts at Southern Methodist University and the Meadows Mental Health Policy Institute, an independent nonprofit organization. Both were in Dallas.

"Our third leg is the environment, and that is you," Esterline told Sansom. "We want to do something big." The goal was to establish an institution that would blend traditional academic activities such as research and publication with those of a nonprofit organization that serves citizens outside academia. This philanthropic mission set it apart from the big water institutes at Texas A&M University and the University of Texas.

In 2012, the Meadows Foundation went big. It gave $5 million in a series of endowments that were designed to attract another $3.75 million in state funds and $1.275 million in private money for a total of $10 million. The institute was renamed The Meadows Center for Water and the Environment.

With the Meadows funding, Sansom aimed to take the center to the next level of influence over environmental policy—just as he had taken Texas Parks and Wildlife and The Nature Conservancy to their next levels. Sansom and Esterline agreed that the center should target its efforts in four areas—research, leadership, education, and stewardship—with policy being the overarching principle.

It was an ideal time to pursue lofty goals in water policy as minds were focused on supply shortages following the 2011 drought in Texas, the worst one-year drought in the state's recorded history. A dozen or more communities were within ninety days of running out of water during 2011, which was the hottest and driest year on record.

While the large infusion of money from the Meadows Foundation catapulted the center into the major league of water institutes in Texas, it also required a delicate balancing act. Sansom had to satisfy two competing interests—Texas State University wanted more publication of scientific research and Meadows wanted more influence over water policy in Texas. Academic publishing was a key part of the university's bid to become a tier-one research institution, while public policy was the "gold standard" for the foundation, which wanted to see lasting impacts that would do "good in Texas."

The balancing act also extended to Sansom's private life. He wanted to spend more time with Nona, who had greater freedom in her schedule following the death of her mother in 2010. For nearly a decade, Nona had been caring for her mother, who had suffered a stroke and traumatic fall shortly after Nona had taken early retirement in 2001. Her mother lived in Houston, requiring much travel and a major time commitment, before moving to Austin.

After her mother's death, Nona began volunteer work for charities and their church, Central Presbyterian in Austin. "Every Thursday morning, we fixed breakfast for our homeless population," Nona explained. "We were in a downtown church and ran a clothes closet." She also donated her time to Manos de Cristo, a charitable group that provides dental care and other services to families in need.

Around 2012, Sansom was traveling extensively in his quest to raise the profile of the Meadows Center, and he wanted Nona to join him on the road. A four-day trip of speechmaking around the state was coming up and he invited Nona to come along. "'It would be fun,'" Nona recollects Andy saying. "Unfortunately, I learned the hard way," she continued. "I went with him. We would race around in the car to get to one event, and he would give his spiel, and then we raced to another event—the same spiel. And then we'd hurry to someplace else. You sit back and wait for everybody to greet him and shake his hand and then pull him away. It was horrible, it was miserable."

Sansom's propensity for connecting with people was both a gift and a curse. "He cares for people," explained Nona. "People would come up to me and say, 'When I'm with Andy, it seems like I'm the only person in the world.' That's exactly the way I felt when we were courting. So I knew what those people were saying. But then when I married them, I realized there were a lot of other people in the world."

As Sansom's job at the Meadows Center became increasingly demanding, the couple's priorities differed. Nona wanted more time for volunteer activities and greater flexibility in her schedule than the previous decade when she was caring for her mother. Andy was busier than ever at the Meadows Center, expanding it and elevating programs with the infusion of money from the Meadows Foundation. Still, Nona and Andy talked to each other about their differences, and then to a counselor, as they had done thirty-five years earlier. "Andy has evolved, he has so evolved," Nona said, with quiet and deep admiration.

By 2012, Sansom was moving in a new direction on water policy at the Meadows Center. He wanted to explore ways to sustain water resources through a stakeholder process that would identify solutions to known obstacles to long-term reliability of supplies. The process was called the Water Grand Challenges Initiative and was supposed to address urgent issues "outside the normal envelope of water policymakers." Sansom brought together influential and diverse stakeholders from legislative, landowner, business, environmental, and philanthropic fields to find "creative solutions for both short-term and long-term challenges that inhibit the sustainability of Texas' water resources for the future."

Funded by the Meadows Foundation and the Cynthia and George Mitchell Foundation, the seven-year effort identified six water goals to guide policy and implementation:

- Providing adequate and sustainable water supplies for all Texans, allowing a thriving future for the state
- Ensuring scientifically sound environmental flows are equally prioritized within an enforceable regulatory framework
- Sustainably managing surface water and groundwater resources to meet current and future needs while avoiding unacceptable impacts
- Investing in water resources to efficiently meet human and environmental needs
- Cultivating awareness and stewardship of vulnerable water resources
- Ensuring all Texas waters are clean, healthy, and life-sustaining

The intent was to advance the long-term goals through short-term actions and theories of change for each issue. Nearly two dozen white papers followed in 2013 alone, identifying critical needs in water policy. The papers were written collaboratively by the Grand Challenge stakeholders and Meadows Center staff members and highlighted areas where regulatory frameworks demanded improvement:

- More precise estimates of water availability
- Better regulation of water flows for the environment
- Joint management of groundwater and surface water
- Recognition of climate change impacts on droughts, floods, and rivers

An overview of the first year's work concluded with a stark warning. It acknowledged that little change was likely despite the policy imperatives for water issues. "Texas maintains an attitude of limited governmental control and individual rights reigning supreme to that of public interest," the report concluded. "Individual management of the most precious resource in the state provides owners with a sense of freedom and liberty. That is to say, 'I possess water on my land that I am responsible for managing and using as I see fit.' But . . . water resource management in Texas is a particularly sensitive subject that needs clarification due to recent shortages and increasing population."

The Water Grand Challenges project issued a final report in 2020 that outlined an action plan for data development, policy analysis, education, funding, and advocacy. While implementation was spotty, Sansom expanded the Meadows Center's expertise in water sustainability by hiring an expert on the subject. Few in Texas were better suited than Carlos Rubinstein. Considered one of the brightest minds involved in state water policy, Rubinstein had chaired the Texas Water Development Board and served as a member of the Texas Commission on Environmental Quality. He grew up in Mexico City in a Jewish family and met Sansom in the late 1990s when Sansom was running Texas Parks and Wildlife. At that time, Rubinstein was the city manager of Brownsville, which was in talks with TPWD, US Fish and Wildlife, and communities in the Rio Grande Valley to create a World Birding Center.

The location of the planned center was a source of competition between the communities, Rubinstein explained. "The legislature was messing with the selection process, and we complained about that during the session," he said. "Ultimately, common sense prevailed and TPWD was allowed to find an acceptable and equitable solution." The World Birding Center was established as a network of birding sites in the Lower Rio Grande Valley. Thereafter, Sansom and Rubinstein remained in touch through their shared zeal for water policy. After Sansom and Rubinstein retired from government service, both worked as consultants and often shared clients on water policy, planning, conservation, and environmental flows projects.

"Andy is a great human being who gives and gives and gives for the common good with a focus on protection of our natural resources," Rubinstein observed.

In 2016, Rubinstein joined the center as a Meadows Fellow, and his first project was overseeing an effort to buy and sell water for the environment through a consortium of conservation groups, including the Meadows Center. The consortium, known as the Texas Environmental Flows Initiative, focused on bays and estuaries where freshwater inflows were imperiled by population growth and increasing demands along the rivers and streams that feed coastal waters.

One of the initiative's earliest and most visible outcomes was the creation of Texas Water Trade, a nonprofit organization that aimed to promote water conservation through market mechanisms. "By creating a market and limiting the amount of water that can be used during certain periods, water becomes really valuable," Sansom explained to podcast host Todd Votteler of the *Talk+Water* podcast. "You manage the resource sustainably so the rights actually reflect their inherent value." Two main obstacles hindered the development of water markets, Sansom explained. The absence of cap-and-trade systems in Texas groundwater management, apart from the Edwards Aquifer Authority, rendered markets unworkable because supply must be limited for trades to occur. The second obstacle was the "prior appropriation" doctrine in Texas law that governs surface water such as rivers and lakes. Prior appropriation means "first come, first served"—holders of older water rights get all their water before junior holders get theirs. The "first in time, first in right" doctrine skews the water supply.

In any case, the Environmental Flows Initiative identified two potential water deals, though both were stalled in 2022. One was the Anahuac transaction, which would have acquired water from a rights holder and moved it to the Anahuac Wildlife Refuge bordering Galveston Bay to restore wetland habitat and increase inflows to the eastern reaches of the bay. The project was put on hold, pending resolution of an unrelated lawsuit between the water rights holder and the state, and later completed by Texas Water Trade.

The second potential deal involved Tres Palacios Bay, an inlet of Matagorda Bay, where preliminary talks were held to buy water rights from a self-declared willing seller. The concept attracted interest from state decision-makers in charge of an oil-spill fund, though talks eventually stalled out.

Water sustainability was sorely tested again in 2015, after Texas suffered its third major drought in four years. Drought was "clearly the most serious natural resource challenge facing coming generations," Sansom wrote in a Meadows Center annual report.

"We wanted to use climate change as the leading part of the conversation," and Sansom was to be "the tip of the lance," Esterline explained. "Are you up to it?" Sansom was asked. He was, and Meadows started giving around $500,000 a year to coordinate research on climate issues and build up climate change expertise within the center. To carry out this mission, the center increasingly sought a collaborative approach. "We take a consortium approach. There aren't that many of us," Esterline said. "We are not about branding so much."

By the time Sansom had been at the center for fifteen years, he began thinking about leadership succession. While Emily Warren seemed like a natural successor, she didn't have a doctorate degree. Sansom himself was required to get one after founding the River Systems Institute.

The center's goals remained in place—research, education, stewardship, and leadership—but they were trickier to balance. The Meadows Foundation wanted to influence water policy and Texas State University wanted more research and academic publishing. In fact, some of the center's proposed work was being nixed by the university because it didn't advance Texas State's goals, according to insiders.

In 2017, Sansom made a presentation to the Texas Water Development Board, where Robert Mace was running the Water Science and Conservation office that studied rivers, aquifers, desalination, rainwater collection, and water reuse. Mace, a nationally known expert in groundwater science, had spent seventeen years at the agency and was contemplating his next move. He'd known Sansom for years, having watched him testify at the capitol, and was aware of his charm and impeccable reputation. "You just can't help but fall in love with him the first time you interact with him," Mace said.

Mace reached out to Sansom after his presentation to the TWDB Board of Directors, where he'd mentioned the Meadows Center's plans for an endowed professorship of groundwater. Mace was intrigued by the possibility of teaching at the university level, having guest lectured at Texas State University and the University of Texas, and explained that he was nearly eligible for state retirement. "That's when the fun discussions started," Mace recalled.

It became clear that Sansom was looking for an heir apparent to the center, in Mace's opinion. That wasn't exactly what he was looking for, especially considering the requirement for academic publishing, which was harder than teaching, in Mace's view. Other aspects of the job gave reason for pause, such as the need to find soft money, or funding that comes from outside the institute or university. Mace had done that at the University of Texas' Bureau of Economic Geology and knew it wasn't easy. The uncertainty of soft money puts people's jobs at risk. Moreover, the center director was under university administration, specifically the provost's office. That meant any visits to legislators, for example, had to be approved.

Still, Mace figured it could be interesting and was hired in 2017 as deputy executive director of the Meadows Center. In the beginning, he thought he was going to be a groundwater professor, though the relationship of the institute to the university was unclear to him. So were the university's structure and the academic lingo. "I had no idea what they were doing or talking about," he said with a laugh.

At that time, Texas State University was moving aggressively with efforts to become a tier-one research university rather than just a teaching college, which historically it had been, and that goal clearly was conveyed to Mace. During Mace's interviews for the deputy position at the center, he was told that research was a top priority. In fact, he got the impression that the center's days were numbered if it didn't transform itself into more of a research entity. Only a small amount of the center's work was being published in peer-reviewed journals.

That Sword of Damocles didn't frighten Mace, a hydrogeologist with a doctorate and a record as a consummate scientist. "I felt like it would be cool to get more into the research side of things again," he said, having spent much of his time on policy at the water development board.

In 2019, Mace was promoted to interim executive director while the university carried out a national search for a permanent one. During the transition period, Sansom kept his corner office at the center, though he was spending less and less time there. By 2020, Mace was selected by the search committee to be the executive director, and he expected Sansom to move out of his office as a signal to the staff of the leadership change. That didn't happen. Emails were exchanged and nothing happened. Mace eventually made an in-person appeal and the handoff finally occurred.

For Sansom, as for many founders, handing over the reins was weighted with myriad emotions—loss, relief, fear, acceptance. He retained the title of founder and continued to teach in the geography department. Moreover, he remained a sought-after consultant and eminence gris in the environmental industry.

Meanwhile, Mace had decided that climate change would be a top priority for the Meadows Center, even though the phrase remained taboo in many government circles and contempt for science was worse than twenty years earlier. At the capitol, "there is no love for the science. It doesn't matter what the science says, the politics trump," Mace admitted.

Up to 85 percent of Texans accept that humans are causing climate change, but they don't believe it's affecting their daily lives. Nevertheless, the impacts are hitting Texas more than most states—in extreme drought, flash flooding, and rising seas. "One thing we have going for us is that it's getting worse every year," Mace said. "It's happening in real time."

Mace had a flair for presenting scientific and technical information in ways that are understandable to a wider audience beyond hydrologists, engineers, and academics. For nearly two decades at the TWDB, he made compelling presentations about climate change without ever using the term, because of agency rules banning its use. "My data is friendly and useful," Mace explained. His Twitter handle read, "I love helping people understand water so they can make informed decisions."

Given state government skepticism about global warming, the Meadows Center would provide the resources that were lacking from the state, particularly practical tools that could help water industry leaders make informed decisions. Mace believed that some water utilities, river authorities, groundwater districts, and legislators were worried about climate change and its impacts on water. "If you're the manager of a water supply system in a very conservative town, you can go into your office, close the door, lock it, and bring up our website to see how reliable or less reliable your surface water rights are," Mace explained. "Depending on the politics of the area, the manager can hide behind terms like 'building resilience,' which can amount to an arbitrary buffer of water supply, or use the center's tool and have greater precision."

The state failure to do research and provide a policy response to climate change was leaving residents, businesses, and the environment exposed to greater risk than necessary. "People don't have the data to know what's coming," Mace warned. "The governor would prefer not to show the impacts of a hotter and drier climate on the state's water supply because it is going to look ugly," he said.

Mace had a clear plan for the Meadows Center to fill the breach. "I want to do these analyses and give them to the people—from the governor down to a farmer, or somebody in an apartment building," he explained. "They can bring up the data on their computer and see what climate change means for them, both in terms of temperature and rainfall. They can also see the resiliency of their surface water supplies and groundwater supplies."

Mace brought to the Meadows a skillset, background, and personality that contrasted sharply with Sansom. Mace was a geophysicist, groundwater modeler, and hydrogeologist who had spent most of his career in state government and academia. He described himself as a "goofball." Sansom was an environmental conservationist and parks and recreation specialist, with a talent for teaching.

"He's totally different from me, and he's better," Sansom said of Mace. "Couldn't be more different, but it's so exciting to watch that happen, and the staff loves him—who wouldn't?"

Chapter 17

Water in a State of Perennial Drought

When Sansom was a child, water amounted to entertainment—swimming, boating, and fishing. When he was a youth, it provided employment, such as lifeguarding. By the time he was an adult, water needed his protection.

For much of Texas' history, water was viewed as a resource to tame and commercialize more than one whose existence determines our own. During the 1970s, the state of Texas took halting steps to safeguard environmental flows. In 1975, the state legislature directed the Texas Water Rights Commission (forerunner of the TCEQ) to consider the impacts of permit applications on bays and estuaries. That legislation also instructed several state agencies, including Texas Parks and Wildlife, to study whether and how freshwater inflows to those coastal waters maintained their ecological environment. Subsequent legislation in the eighties and nineties arrived at the term "beneficial inflows" to describe enough water draining into bays and estuaries to maintain sport and commercial fishing and estuarine life, focusing only on minimum amounts needed to sustain aquatic life.

As an environmental conservation leader in the 1980s, Sansom was able to protect water in selected places. On a statewide scale, however, he was only able to preserve the resource when he took the helm of Texas Parks and Wildlife Department in 1990 and could set policy. Having spent years assessing the fragility of ecosystems for The Nature Conservancy, Sansom recognized the critical nature of environmental flows, which included the amount of water needed in a stream for a sound ecosystem. Balancing the health of rivers with the human demands of a fast-growing population was a raison d'etre of Sansom.

A major obstacle was that water for the environment is not legally recognized as a "beneficial" use of the resource, which is a requirement for permitting purposes. Beneficial uses include domestic, agriculture, industrial, and power—but not fish and wildlife. Water that wasn't deemed of benefit to humans was considered waste. In 1985, the legislature made a gesture toward protecting rivers and streams through the permitting process, though the focus remained on ensuring bay and estuary health for commercial fishing. The 1985 legislation required the environmental conditions of bays and estuaries be considered by the Texas Water Rights Commission in future permitting, though the practical effect was limited, given that most of the state's water already had been allocated with no provision for plants and wildlife. Very little water was left that could be permitted for environmental purposes.

The legislation followed a multiyear drought that left some towns in Texas without drinking water and reminded state lawmakers that not only do people need reliable supplies of water, but so do rivers and bays. In 1984, the Lavaca River was providing too little water to Lavaca Bay, which empties into Matagorda Bay, to keep salinity at healthy levels for the shrimping industry. That prompted a state-ordered release of freshwater from the river to lower salinity. Economic impacts got lawmakers' attention more quickly than environmental ones, though the realization was dawning that the two were intimately connected.

Still, the idea that water was wasted if it didn't directly benefit people prevailed during most of the period 1987 to 2020 when hydrologist Cindy Loeffler worked for Parks and Wildlife as a leader in various water capacities, often engaging with landowners and water managers. "Individuals as well as water officials would talk about water being wasted," she recalled about her early years.

In 2022, the sentiment lingered. "In the legislature over the last three or four years, I've heard members in committee hearings say things like, 'The last drop of water that goes over the last dam into the Gulf of Mexico is wasted,'" Sansom told Dr. Jen C. Brown, an associate professor of history at Texas A&M University-Corpus Christi. "That is still a belief among some members of the legislature."

In fact, legislative support for environmental protection has waned over forty years, according to Sansom. "If you look back at the period of time during the mid-1980s, the legislature was much more environmentally inclined than any time since," Sansom told Brown, an environmental and oral historian who writes, produces, and narrates

The Gulf Podcast. "The legislature was critical of the department [Texas Parks and Wildlife] for not being aggressive enough in protecting the environment. The department did a great job of managing white-tailed deer or running state parks, but in terms of being actively involved in environmental protection, we were not."

In response to the legislative criticism, TPWD created a Natural Resource Protection Division and invited environmental scientist Susan Rieff to lead it. Rieff's job was to research and make recommendations on freshwater inflows to estuaries and instream flows for rivers and reservoirs, as well as for the benefit of wetland conservation and restoration. The intent was to recommend how much water should flow in streams to protect fish and wildlife while also benefiting human uses of water. "If somebody went to apply for a big water right on the Guadalupe [River], then the department was a party to that discussion," Sansom explained. "It was controversial because every one of those transactions was controversial."

When Sansom took over as CEO of Parks and Wildlife in 1990, he and Rieff's successor, Larry McKinney, worked together closely. "Larry McKinney used to come in my office almost every day [and] the two of us would check to see which/if one of us had been fired," Sansom recounted to Brown. "That's how difficult those kinds of things have been in Texas."

The pioneering work with McKinney revealed a major shift in perception for Sansom. "I had the most significant epiphany as to the significance of water to almost everything else I was involved in—from protection of biodiversity to opportunities for outdoor recreation, including hunting and fishing and paddling," Sansom said. "It became apparent to me during the 1990s that Texas has probably the most intact system of bays and estuaries of any state, but they're entirely dependent on the continued flow of freshwater down to those wetlands and estuaries, and we still don't do a whole lot to protect [them]."

The revelation prompted Sansom to take another pioneering step. In 1997, he created a water resources program at Parks and Wildlife that would view the resource as a necessary component of the environment, rather than simply recreation for fishing and hunting enthusiasts. The program was intended to provide a platform from which TPWD could work with the Texas Commission on Environmental Quality and Texas Water Development Board on a peer level.

Sansom staffed the Water Resources Branch with talented scientists such as Loeffler, the hydrologist who specialized in instream flows and had started work at the agency in 1987. She understood how physical and hydrological changes impacted fish and wildlife, including those in bays and estuaries. Equally important, she possessed a pleasant demeanor that served her well in presenting Parks and Wildlife's views in conversations with interagency collaborators, landowners, and individual stakeholders.

A milestone in Texas environmental law came in 1994 with publication of a seminal report entitled "Freshwater Inflows for Texas Bays and Estuaries." Signed by Sansom and his counterparts at TCEQ and TWDB, the report was the culmination of 1985 legislation that required environmental conditions to be considered in future water rights for the purpose of protecting instream flows. The report provided scientific grounding for regulating the amounts and timing of freshwater flows needed to protect Texas estuaries.

"This was a real turning point," said Loeffler, who spent more than thirty years at the agency. "TPWD became much more involved and outspoken regarding water for fish and wildlife. We had a seat at the table." The agency used the study results to make recommendations on applications for water permits that would set precedents for years to come. These included the Lower Colorado River Authority's Highland Lakes and the City of Corpus Christi's Choke Canyon Reservoir.

"Water rights are often quite contentious, and these were no exception," Loeffler recalled. "We had Andy's support to be fully engaged in these deliberations."

Thereafter, TPWD's role in water governance continued to expand. In 1997, legislation that created a state water-planning process gave the agency a nonvoting seat in each of the sixteen regional groups tasked with developing the plan. "We heard things like, 'We don't have the luxury of considering the environment,'" Loeffler recounted. The disdain did not deter the agency.

In fact, TPWD's comments on the first state water plan "really got folks' attention," Loeffler noted. "We took a pretty strong position" on the role the agency should play in designating river or stream segments as "ecologically significant," meaning that no state funds could be used for building a dam on that segment. TPWD identified specific criteria to be used by regional water planning groups in recommending stream segments to the state legislature, which ultimately made the designation.

"Andy took a lot of heat for that letter, but it reinforced the importance of not just considering fish and wildlife, but actually planning for them," Loeffler explained. Sansom's defense of vulnerable stream segments may have antagonized his political foes.

While at Parks and Wildlife, Sansom used his flair for branding and advertising to carry out a ten-year educational campaign called "Texas: The State of Water" to explain how to protect water resources and wildlife. The campaign included a series of *Texas Parks & Wildlife Magazine* issues and documentary videos that explored water issues in depth. The first documentary was narrated by Walter Cronkite, who had studied journalism at the University of Texas at Austin and worked as a reporter in Houston. He was the first prominent journalist of his time to recognize the importance of the environment as a news story. The TPWD campaign was aimed at a wide audience ranging from the public to lawmakers to students, and it helped raise the profile of the agency beyond hunters and fishermen.

"This campaign helped enlighten folks in the water community and changed thinking," according to Loeffler. The belief that water flowing across the countryside to the coast was a wasted resource started waning.

Sansom's biggest opportunity for shaping water policy came after he moved to Texas State University's International Institute for Sustainable Water Resources in 2002. The regulatory framework for instream flows was evolving with input from state agencies, academic institutions, and environmental groups—and Sansom wanted the institute to be a player in both instream and environmental flows. His goal was to establish a collaborative effort with bigger entities through his extensive network of industry contacts, carving out a role for his smaller and newer institute amid the larger and more experienced parties that could do extensive research and field work. He brought Pulich to the institute to help him carry out the mission, and Pulich became the "father of e-flows."

"My priority was to do everything we could to make sure that we were developing those inflow numbers with precision," Pulich said. A research biologist, he could see that it would take five years to identify the ideal amounts and timing of inflows into each bay on the Texas coast. "We finally figured out the process, the methodology with the

Water Development Board. The first really good set of numbers for the Guadalupe Estuary became the template for the rest of the estuaries. We used that same methodology to do another six."

Pulich's pioneering work in environmental flows continued after he was brought to the River Systems Institute by Sansom. Pulich's deep knowledge of seagrass, which served as a kind of sentinel species for bay and estuary habitats, informed his environmental flows work and drove him to seek accurate numbers for the flow regimes so they could be converted into governing rules.

While Sansom saw environmental flows as a major policy objective, he also had to focus on the operational aspects of the institute that often diverted his attention. Winning new grant funding was a constant concern, while keeping and recruiting new staff was essential to building the institute's profile in the university system. The major strategic shift in 2005 that changed the name to River Systems Institute demanded rebranding, money, and resources. The competing demands on Sansom's attention diluted his focus on environmental flows.

The stakes rose considerably in 2007 when the state legislature passed Senate Bill 3, which created a pioneering process for environmental flows regulation, driven by stakeholders and scientists. The process was designed to answer three questions:

- How much water is needed to sustain a sound ecological environment in the state's rivers and estuaries?
- How can this water be protected?
- What is the appropriate balance between water needed to sustain a sound ecological environment and water needed for human or other uses?

The answers were to come from stakeholder committees and science advisory panels appointed in each of seven river basins that feed into the state's major estuaries. The stakeholders and scientists were charged with making recommendations to an overarching Environmental Flows Advisory Group. In turn, the group made recommendations to TCEQ to be converted into standards. The elaborate process designated categories of stakeholders—environmental, economic, agriculture, industry, and municipal interests—who would review recommendations from expert science teams. Present and future water needs were considered.

By 2014, the process proved to be deeply flawed and disappointing to many. Very little water was available for environmental purposes because most already had been allocated for human purposes via water rights permits. Most of the standards set by TCEQ, known as a reluctant regulator, were so low that they had little effect on new water rights and virtually no effect on existing ones. Critically, TCEQ showed little interest in sustained support for e-flow standards, ignoring a mandate to reevaluate e-flow standards every ten years.

"TCEQ is the agency of broken dreams," observed Dan Opdyke, an environmental engineer with a doctorate degree who formerly worked for Parks and Wildlife and often engaged with the environmental commission. "Stasis will likely continue as environmental advocates are hard-pressed to push for change because the existence of e-flow standards gives the impression to many that the environment is being protected."

A decade later, in 2022, TCEQ's negligence toward environmental flows came under unusually blunt criticism, echoing some of the regulatory weaknesses that Sansom had seen ten years earlier. The criticism of TCEQ came from the Sunset Commission of Texas, which periodically reviews state agencies to determine whether they should continue in existence or be abolished. TCEQ was deemed worthy of continued existence, but it was faulted for "an unclear statutory framework [that] has stalled the state's process for developing environmental flow standards . . . leaving participants unsure how to proceed with adopting and updating flow standards for the state's river basins and bays."

The Sunset Commission recommended that the Environmental Flows Advisory Group and science panel resume work. More importantly, the commission slammed TCEQ for failing to set aside even the small amount of water available for environmental protection. The agency was chastised for neglecting to revoke water rights that weren't used for long periods—sometimes decades—and reallocating them for the environment.

The commission recommendations were considered by the legislature during its 2023 regular session. Sansom acknowledged some very modest improvements in statute, but he remained alarmed that lawmakers still didn't seem to consider environmental flows to Texas' bays and estuaries to be a very important issue. "As a result, our coastal resources continue to be at risk," Sansom lamented.

Sansom's most flamboyant bid to influence policy was in 2018 when he sought to create a water awareness program together with the TWDB and advertising legend Roy Spence, cofounder of the GSD&M advertising agency and creator of the iconic "Don't Mess With Texas" campaign. Spence had been approached by Jeff Walker, executive administrator of the Texas Water Development Board, who knew him from their childhood growing up together in Brownwood. Sansom's fascination with advertising went all the way back to his time at the Department of Energy when he worked on the "Don't be Fuelish" advertising campaign.

Known for his theatrics, Spence gave a glossy presentation to a TWDB water conference, saying his newfound love for the resource stemmed from reading the state water plan all in one night. "If you don't let the people of Texas know what's at stake, you're cheating them of the opportunity to make a better Texas," he told the high-profile conference. "Can I help y'all?" he quipped. "I wrote a theme line: 'It's Texas, dammit. We've got oil and we've got gas and we've got technology and we've got people. We've got to have water so we can kick butt forever.'"

The match seemed made in heaven, given Sansom's appreciation of marketing and Spence's abiding thrill in the "hunt for sea-changing ideas," as the Advertising Hall of Fame put it. The Meadows Center provided $84,000 for preliminary research and a creative development campaign, with support from the Ewing Halsell Foundation. "What we really wanted to do was to create a 'Don't Mess With Texas' for water," Sansom explained, referring to the thirty-five-year-old public awareness campaign to combat litter in the state. The team came up with a slogan, "Texas—Do or Dry," modeled on the call to action in "Don't Mess With Texas." GSD&M developed graphics and other collateral materials.

The miscalculation that Sansom and Walker made was they didn't think that the Texas Water Development Board would balk. The agency declined to fund the Spence campaign, perhaps worried about the appearance of using taxpayer dollars for glossy advertising. The refusal led Sansom and other water leaders to turn to the Texas Water Foundation, a preeminent nonprofit that aimed to raise awareness of water and promote leadership in the industry. The foundation, led by its new

CEO, Sarah Schlessinger, ran with the idea and developed a campaign around the tagline "Texas Runs on Water." The foundation partnered with local entities to adapt and publicize the message—for example the city of Houston and "Houston Runs on Water." In contrast to the Spence campaign, the TWDB was willing to support the Texas Water Foundation effort.

"I was a little bit surprised at that because what Jeff and I were looking for was more of a call to action," Sansom said. "But apparently, it's been well received and is being piloted around the state."

Looking ahead, the biggest challenge for water was sustainability—ensuring enough for all living things well into the future. Conservation is the starting point, and Sansom has long been an advocate for efficient use, particularly in the agricultural and municipal sectors. Water and conservation were the focus of several of the nine books he wrote and edited.

Recycling sewage also held promise, he noted, though its impacts needed to be carefully considered. Reusing treated wastewater instead of releasing it into rivers could reduce the amount of water available for those living downstream, he explained in the 2017 book *Of Texas Rivers and Texas Art*. As someone who grew up in the bottom reach of the Brazos River, Sansom was always attuned to those living downstream. "As pressure to provide more water for municipal growth . . . intensifies," he wrote, "increasing efforts will be made to 'reuse' treated wastewater, potentially reducing its release into the river and, in turn, threatening the welfare of those who live, work, and play downstream."

While conservation and recycling were necessary, Sansom believed the most promising method of sustaining water supplies was the private market. A market had proven successful in the eight counties comprising the Edwards Aquifer Authority, a groundwater district that created a cap-and-trade system in 1997.

"If I were a dictator, what I would do is establish caps within the jurisdiction of the groundwater districts like on the Edwards," Sansom mused aloud to Dr. Jen Brown of the Texas A&M University-Corpus Christi oral history project. "First and foremost, that allows for our market to take place. There's potential funding for purchases of groundwater to protect that resource, but why would anyone do it when across the property line the neighbor can drill a well and just pump it dry anyway.

So there's no marketplace there for that reason. I think we'll find some greater movement toward a cap so that organizations like The Nature Conservancy and Texas Water Trade can do deals to protect water in a market, in a free market."

The downside of water markets was that agriculture often suffered. When farmers using irrigation don't farm because they've leased or sold their water rights, their payroll shrinks, the agricultural infrastructure declines, and their land can lose value. "Water markets across the United States have commonly facilitated the sale of mainly agricultural water largely to nonagricultural users," according to a doctoral dissertation by Anastasia Thayer at Texas A&M University in 2018. "Sales have led to decline in agricultural production and have induced regional losses within the rural economy." In the Edwards Aquifer Authority, "increased water market transfers negatively affected the agricultural industry as captured through changes in agricultural payroll."

Sansom believed the risks of water markets could be mitigated. "Often, the 'forbearance' is temporary, and farmers can get cash for reducing consumption at a time when crop prices are down," he contended. "Pulling back to a less water intensive crop makes sense; or other factors make reduction attractive. In the case of permanent water rights purchase, normally the land has already gone out of agriculture for other reasons."

Overarching all water policy, however, was climate change. "Knowing the uncertainties of global climate change will help us prepare more thoughtfully for the future," Sansom wrote in his 2008 book, *Water in Texas.* That uncertainty poses a clear risk for water planning in Texas, which is based on the drought of the 1950s. Future droughts are projected to be worse and require rethinking.

Given the universal need for water, sustainability was essential. All too often, sustainability was framed as a choice between people and environment. This was a false choice, in Sansom's opinion. Water security and ecological integrity were tied together, as the World Water Council put it. Public policy needed to ensure that water management incorporates economic development, disaster resilience, and biodiversity preservation.

To make that happen in Texas, water law needed to evolve to treat rivers and aquifers as connected water resources. "To continue to pretend that aquifer use . . . is a different issue from withdrawal of water from

. . . streams . . . is to risk major damage to our river systems," Sansom wrote in *Water in Texas*. "Policymakers must push for conjunctive use of these resources, treating them as they actually are: the same."

Students studying Sansom's books were more likely to change policy in the future than lawmakers of his time were to revise it any time soon.

Chapter 18
Learn or Languish

In 2005, American author Richard Louv wrote a book about the human costs of spending too little time in nature—maladies such as physical and emotional illness, attention difficulties, sensory loss, and obesity. In his book *Last Child in the Woods*, Louv coined the term "nature-deficit disorder" to describe these ailments. He argued that children, in particular, should spend time in natural surroundings to build confidence, serenity, and focus.

Louv's work resonated with Sansom, whose childhood enjoyment of the outdoors evolved into student activism for Earth Day and later into teaching about environmental leadership. He had long sought to get kids into parks and fresh air all the way back to high school when he taught swimming at a community pool. As an adult, Sansom continued to promote greater access to green spaces for youngsters, especially those living in cities. The more children knew about the natural environment, the more they would want to be in it, according to Sansom.

In the 1960s, elementary and high school education about the environment usually was embedded in science classes. Curriculum typically focused on nature, the outdoors, and conservation, according to Bora Simmons, professor of teacher education at Northern Illinois University. Science projects and fairs were designed to give students the opportunity choose a topic, research it, and give a report.

The year 1970 marked a turning point in environmental education. The first Earth Day raised awareness of the need to formalize instruction and take a more interdisciplinary approach that spanned biological, physical, and chemical interactions in the world. It wasn't until the 1990s, though, that the interdisciplinary approach really gained traction. Educators pushed for standardized curricula in schools, and environmental advocates hoped to influence a generation of students who would take those values into adulthood, shaping public policy. That same year Congress passed landmark legislation creating the Environmental Protection Agency and the Clean Water Act.

The EPA itself defines environmental education as:

- Awareness and sensitivity to the environment and environmental challenges
- Knowledge and understanding of the environment and environmental challenges
- Attitudes of concern for the environment and motivation to improve or maintain environmental quality
- Skills to identify and help resolve environmental challenges
- Participation in activities that lead to the resolution of environmental challenges

An environmental education is "a process that allows individuals to explore environmental issues, engage in problem solving, and take action to improve the environment. As a result, individuals develop a deeper understanding of environmental issues and have the skills to make informed and responsible decisions," according to the EPA. The process teaches critical thinking.

Educators made various efforts to improve environmental curriculum and teaching practices to reach students more effectively. By 1990, Congress stepped in to create the National Environmental Education Foundation to cultivate an environmental consciousness and responsible public.

Over the ensuring years, research demonstrated the benefits of environmental education, including improved academic achievement, greater environmental stewardship, deeper personal wellbeing, and stronger communities. "Make the environment more accessible, relatable, relevant, and connected to the daily lives of all Americans," the National Environmental Education Foundation urged. In addition, support grew for environmental studies as a pathway to science, technology, engineering, and mathematics careers.

Sansom's view of environmental education was that it needed to be modernized to make it more relevant to a wider audience, thereby increasing appreciation of nature. In 2005, a kindred spirit resurfaced. Sansom had met Molly Stevens in the 1990s when she was a fundraiser for The Nature Conservancy, shortly after he had left the organization. They stayed in touch while Sansom was at Parks and Wildlife and she moved to the Environmental Defense Fund. In the mid-aughts, they reconnected over a common goal—getting children into nature.

Stevens had just started work at the Westcave Outdoor Discovery Center, a nonprofit that offered educational hikes and activities around a magnificent waterfall and grotto in a limestone sinkhole situated in an arid savannah near Austin. The center promoted enjoyment and protection of nature through education and conservation for all ages. Westcave was a trailblazer in Texas and beyond in developing ways for children to enjoy nature and learn about the environment. Not only did the curriculum align with the state's standardized testing, it also provided a uniquely tactile experience for youngsters.

"It's really mostly the sensuousness of being in a place like Westcave," Stevens told David Todd of the Texas Legacy Project. "What does it smell like? What does it look like? What does it sound like? What do you not hear? What does the bark of this tree feel like—the awakening all of the senses.

"Twice during the tour, they stop and are just quiet. We ask them not to shuffle their feet, not to speak, just pay some attention to what they hear. Oftentimes, we hear back from kids that that was their favorite part of the day—that time to be really quiet, close their eyes, and listen to the birds, listen to the way the wind sounds and the leaves."

Westcave led a drive to create the Children In Nature Collaborative of Austin, a grassroots group that became part of the Texas Children in Nature Network, a six hundred-partner coalition that connects "children and families with nature in Texas to be healthier, happier, and smarter."

The concerted efforts made Texas the top state in the country for getting children outdoors, said Stevens. The momentum dovetailed well with Sansom's view that better education about the environment would inspire a greater appreciation of it. Not long afterward, he joined Westcave's board of directors.

By the mid-aughts, Sansom was in a position to implement his evolving ideas on environmental education. As the head of the Meadows Center for Water and the Environment, one of his charges was to educate audiences ranging from young pupils to university students and interested citizens. Education was one of the four goals that the Meadows Foundation had for the center, along with research, leadership, and stewardship.

"One of the most important roles he played was a cheerleader for the whole movement," Stevens explained. "Every chance he got he would talk about the profound negative impact of the disconnect from nature and the importance of finding ways to get all children out into nature. It was in his DNA to advocate for children to have significant regular opportunities to play and learn in nature."

Sansom and Stevens had breakfast regularly at the Magnolia Cafe in West Austin, the neohippie bistro that had served as Sansom's unofficial morning office for twenty years. He networked, negotiated, mentored, and once fired an employee in the eatery, a place that felt like a treehouse with big, midcentury modern windows in one of the two dining rooms.

At one of the breakfasts, Stevens fretted to Sansom that Westcave couldn't meet a loan payment. "Let's just get the best minds in Austin together and figure out what we're gonna do," Sansom responded, according to Stevens. He pulled together leaders of The Nature Conservancy, Parks and Wildlife, and LCRA, as well as donors, landowners, and himself into "the most impressive room of people I was in during my years at Westcave," Stevens recollected. By the end of the meeting, LCRA agreed to buy additional land around the Westcave Preserve, thereby paying off the loan.

"That was Andy, he was a problem solver," Stevens said.

A growing number of studies were showing children spending less time in nature, often with detrimental effects, and often due to more time spent indoors on digital devices. Scholars were clutching their pearls, and even Bill Gates limited his children's screen time.

In 2012, Sansom had an epiphany about technology and nature, during a Texas Tribune Festival event where he chaired a panel discussion. "I realized that technology could be an important, if not essential, tool in environmental education," he said. "Each of the panelists bemoaned the fact that our children were spending so much time in front of electronic screens and not enough time outdoors. A young women of high school age challenged the panel by stating that if we were going to expect young people to make changes to help protect the environment, we are going to have to use technology."

A light bulb went on for Sansom. "We began almost immediately to include technology in our environmental education program at The Meadows Center. First, we put QR codes around the site so that visitors could get additional information about what they were seeing on their cell phones. We then pioneered the use of virtual reality technology by allowing students to experience scuba diving in the lake without ever getting wet."

Stevens was coming to the same conclusion. "We have to meet people where they are," she explained. "Technology is not just for playing games, but for doing research and writing their papers." She gave credit to Sansom for being "a real hero for Westcave."

In 2015, Westcave sponsored a conference on technology in nature, as suggested by Sansom. The Meadows Foundation provided generous funding. The conference showcased popular apps for learning about nature, such as the Audubon Bird Guide app.

The creative use of technology fit in with Sansom's philosophy of education. He had started teaching environmental classes when he arrived at Texas State University in 2001 to run the International Institute for Sustainable Water Resources. He taught in the Department of Geography and Environmental Studies, one of the largest undergraduate programs of its kind in the United States. His courses included "Theory and Practice of Parks and Protected Areas" and "Conservation Leadership." The leadership class introduced students to the conservation movement and philosophy, and it examined problems and attributes of leadership at nongovernmental, local, state, and federal levels. Topics included operational implications, ethical issues, and other considerations for successful leadership.

Sansom's interest in environmental education was particularly focused on informal methods of teaching, the topic of his doctoral dissertation in 2013. Formal methods impart knowledge through direct instruction that is more rigid and resistant to change. Informal education is less structured and even "messy," placing students in settings where they can experiment or tinker in an environment that is more relaxed than the classroom. Sansom examined these contrasting approaches in his dissertation, comparing "formal" and "informal" water education programs at the university, earning a doctorate in geographic education.

In the informal process, students find meaning in addition to facts, Sansom concluded. Activities such as field trips or outdoor experiences create authentic learning experiences, enhancing formal education by giving greater context. In field trips to museums, younger students found meaning when they were allowed to have brief play periods and free choices of exhibits for closer study. These unstructured periods supplemented structured ones such as workbook exercises.

"Rich authentic situations that occur in communities of practice and experience cannot be simulated in the classroom setting," Sansom wrote. Informal learning often does not contain specific outcomes that are known or predicted at the outset, Sansom noted. Moreover, informal approaches can change more rapidly, incorporating new information and methods more quickly. Despite the advantage of rapid adaptation and innovation, most states have not consistently encouraged informal programs, particularly in the areas of science, technology, engineering, and mathematics, he found.

Informal education programs at Texas State University's Spring Lake could achieve the meaning and context needed for environmental literacy, Sansom believed. These opportunities could be enhanced by collaborating with museums, nature centers, summer camps, and nearby preserves, such as Westcave Discovery Center and Jacob's Well Natural Area.

His informal approach resonated with the more advanced students in Sansom's classes. Most of his classes were "stacked," meaning they included both undergraduate and graduate students. Many of those seeking master's or doctorate degrees under Sansom's sponsorship were older students returning to school after working as a professional. High demand, fueled by word of mouth, often meant waiting lists for his courses. "People who want this this type of education seek it out," observed Kevin Colgan, who studied geography under Sansom after serving in the US Navy.

"He had a way of having us students hang on every word," explained Colgan, who was thirty years old at the time. "He's very calm, very quiet. He's not going to waste time delivering an overabundance of opinion. He could sit there and state one sentence that summed up the whole thing."

Sansom was a good storyteller who could "relay his emotion into the story and instill the minutiae of things that caused him to feel the way he did, and then deliver it in a way that built the same emotion for you," said Colgan, who subsequently went to work for one of the largest river authorities in Texas, the Lower Colorado River Authority. Sansom's teaching method was inspirational. "He motivated us to leave the class and become involved. I wanted to be part of saving places."

Sansom's method of motivating students impressed Sharon Savage, another former student. "He seemed so committed to giving back to his students what he knew about the earth so that each of us could find a responsible path forward, hopefully one in favor of the earth," said Savage, who returned to university for a master's degree in environmental sustainability after practicing as an attorney. "He offered everything he knew back to his students to help them." Sansom's holistic approach also resonated with Savage. "He had a way of understanding relationships of humans to each other and to the earth."

One of Sansom's greatest gifts was his ability to influence perceptions, Savage believed. Sansom himself was influenced by Wendell Berry, a twentieth century American poet, environmental activist, and farmer. In Berry's 1970 essay *Discipline and Hope*, he wrote that "ecology may well find its proper disciplines in the arts, whose function is to refine and enliven perception, for ecological principle, however publicly approved, can be enacted only upon the basis of each man's perception of his relation to the world." For Savage, "Dr. Berry set into motion the idea that artists are the receivers of and communicators of the will of Earth."

Savage saw Berry's ideas reflected in Sansom. "Andy's incredible achievement in teaching came from his vast understanding of perceptions and relationships," she explained. "He worked on our perceptions, he ignited an awareness of them and the power within them." She quoted Sansom as saying, "Through art, people understand the story."

Sansom used an artistic lens to examine nature in his book *Of Texas Rivers and Texas Art*, which he coauthored with art collector and educator William E. Reaves. The authors explored riverine themes and art history through oil paintings, watercolors, lithographs, and other artistic works. In his 2018 book *Seasons of Selah*, Sansom took a poetic view of the Bamberger ranch by weaving together photographs and personal musings.

Sansom teamed with Linda J. Reaves to compile *The Art of Texas State Parks: A Centennial Celebration, 1923–2023*, a survey of parks seen through the eyes of artists to commemorate the one hundredth anniversary of the state park system. The book reflected Sansom's view that landscape art plays a vital role in conservation, a belief that he lives out. The walls of his home are adorned with landscape art.

While Sansom was revered by many students, some occasionally felt frustrated by the vagueness of his assignments. "He never gave you a rubric," explained Colgan, noting that younger students missed the specificity more than he did. "There's never a word count, there's never a page count. There's never one specific question that you're answering. He'll say to visit a park or protected place and make a presentation on the challenges. The students pull their hair out because they don't know what he wants."

Sansom also influenced his daughter's work at the Bamberger Ranch Preserve, where she continued the hands-on approach to science classes and nature camps that had been developed by Bamberger's second wife, Margaret. In 2020, April Sansom became the executive director of the preserve and was responsible for carrying out its mission of land stewardship education, among other duties. Educational programs at the preserve taught the importance of biodiversity, the Texas Hill Country's unique habitat and species, and the natural history of Central Texas through participatory experiences aimed at school groups.

"It's a very carefully developed curriculum, and those students actually perform better on the science questions of the STAAR tests," April explained, referring to the State of Texas Assessments of Academic Readiness. "We have data showing that, and we are working to collect that data because we feel it tells an extraordinarily important story."

While Sansom focused on the *approach* to environmental education in his dissertation, the *substance* of that education became increasingly politicized in the first part of the twenty-first century. Local school boards in Texas and elsewhere increasingly reflected the national agenda of political parties. The Republican Party historically saw education as a key driver of economic growth, though that view began to conflict with a growing suspicion of educated elite in the early twenty-first century. Hostility to experts sometimes included contempt for scientific inquiry, which constrained classroom teaching and subjected curriculum and books to scrutiny for cultural biases rather than educational excellence.

"If you're completely driven by ideology, then logic is a threat," Sansom observed. He noted that Robert Mace, his successor at the Meadows Center, was bitterly attacked for a study he conducted on Comanche

Springs in Fort Stockton. "There was a harsh reaction to his scientific findings with respect to the groundwater in the Fort Stockton area. He was threatened, they went after him personally. He was afraid that it was going to affect everything from his own standing in the university to funding for the Meadows Center."

The Mace study examined how much groundwater pumping would have to be cut back to restore Comanche Springs, which dried up in 1961 due to heavy pumping by the Clayton Williams family and other irrigators in the area. In recent years, the springs have flowed for a couple of months a year. The study was done by the Meadows Center in conjunction with Texas Water Trade to determine how voluntary cutbacks in pumping might work.

The attacks on the study came from the Williams, a politically connected oil and gas family that also farmed irrigated crops with groundwater. The family had argued in a 1954 court case that it owned the groundwater under its land and could use any amount it wanted, regardless of the impact on the springs.

Ideology was having a chilling effect on environmental education, according to some observers. The National Center for Science Education and the Texas Freedom Network Education Fund gave Texas a failing grade in 2020 for teaching public school students about climate change. The watchdog groups found the state's science education standards didn't meet key, basic criteria about climate education—namely that climate change is real and is caused by humans burning fossil fuels, and that its effects will be catastrophic.

In her book *Miseducation: How Climate Change is Taught in America*, author Katie Worth described how students in the United States were being misled on climate science because global warming is denied by conservative ideologues on school boards, politically motivated teachers, and some oil companies. In Texas, textbooks raised doubt about the widespread scientific consensus on climate change resulting from human activities, insinuating it is unsettled science. Still, some students learned about the subject in advanced placement classes, preparing them for university education and better jobs in ways not available to others, according to *Miseducation*.

The result was a two-tier system of environmental curriculum in Texas that favored advanced students and penalized the rest, Worth contended.

The two-tier system also put Texas at a competitive disadvantage with other states in educating citizens for better-paying jobs.

"School boards are dictating that you can't use the 'climate' word or 'global warming, ' so Texas kids are not going to have the opportunity that others do," Stevens bewailed. The risk was that Texas would produce second-class students because they weren't learning climate science, she added. Another risk was that students in rural areas, where school boards typically are more conservative than urban ones, would lag behind.

Some teachers were willing and able to teach the curriculum because they were more experienced, better paid, and felt safer in covering the subject, Worth noted. Stevens echoed that observation. "I know a number of science teachers because of the work I've done," she said. "I know that they're teaching climate science in their classrooms They may be calling it something else. But they're making sure that the kids in their classrooms have opportunities to understand what's happening."

Sansom wanted Texas State University to produce environmentally literate citizens and a more water literate population generally. Yet environmental literacy was being politicized at the very time that ecological challenges were escalating—global warming, species extinction, and rising sea levels.

The need for better education about the environment would only grow—and Sansom knew it.

Chapter 19

Creating a Sustainable Ranch

In 1995, Sansom started to write his first book, *Texas Lost: Vanishing Heritage*, a gift to himself for his fiftieth birthday. Writing was a pursuit he had relished for years, though day jobs often got in the way.

When word of the book project reached Terry Hershey, one of his commissioners at TPWD, she suggested to Sansom that he use her ranch near Stonewall as a writer's retreat. As one of the longest-serving members of the commission, Hershey and her husband had hosted Andy and Nona at their 1,560-acre ranch a number of times. Now the offer was to spend time there regularly.

From a prominent Fort Worth family, Hershey attended Stephens College in Columbia, Missouri, earning a degree in philosophy. Her first marriage ended in divorce over differences about having children. Hershey wanted no children as she fervently believed in zero population growth. In 1958, she married Jacob W. "Jake" Hershey, a distant relative of the Hershey chocolate dynasty who had moved from Pennsylvania to Houston in the 1940s after graduating from Yale University and surviving World War II in the South Pacific. He worked in corporate procurement in the oil industry, and one day in the early 1940s, met with a barge line salesman who tried to bribe him. Jake Hershey reported the attempted graft to the owner of the barge line, who proceeded to make Jake head of the company. It ultimately became the American Commercial Barge Line, the largest of its kind in the world.

During the 1960s, Terry Hershey had met Lady Bird Johnson through their mutual friend and colleague, Liz Carpenter, the First Lady's press secretary. Hershey and Johnson hit it off as they discovered their shared commitment to protecting nature, leading to Hershey's membership on the First Lady's Committee on Highway Beautification—which gave rise to the Lady Bird Johnson Wildflower Center in Austin.

In 1969, the Johnsons moved back to their native Texas from Washington, DC, shortly after attending Richard Nixon's presidential inauguration, and the couple spent more time than ever at their ranch near Stonewall in Gillespie County. Sansom thinks Johnson called Hershey to let her know that a nearby ranch had come on the market, and in 1976, Hershey and her husband bought it. The Hershey Ranch was only a few miles away from the Johnson spread, also in Gillespie County. Hershey adored the ranch for its wildlife and natural surroundings.

She would bring friends to relax and enjoy the change of scenery from Houston, and on one occasion she called Lady Bird Johnson to invite her to join the group. Johnson accepted the invitation and arrived shortly thereafter in a black limousine, complete with Secret Service detail. "Terry claimed she had a plate full of sandwiches that she offered to the Secret Service agents, saying, 'Here, take this, and go upstairs and watch TV. We'll take care of Lady Bird,'" according to Nona.

When Hershey joined the Parks and Wildlife Commission, she made no secret of her support for animal rights and opposition to the killing of any creatures, noting that the ranch was "to protect the critters." After her controversial inaugural meeting, she became a tutee of Sansom, who explained the agency's roles and responsibilities. When he talked about animal culling to prevent overpopulation, "she clearly had an epiphany," Sansom recalled. "Just as you have too many people, you can have too many animals. She had a certain philosophical framework, and it was a paradigm shift."

Starting in 1995, the Sansoms were the primary users of the Hershey Ranch and its homestead. They would arrive on Thursdays and stay until Sunday, with Andy writing and Nona reading chapters as he finished them. Sansom had organized his work at Parks and Wildlife so that he could be out of the office on Fridays, a considerable feat in the 1990s when a landline telephone was the only connection to work and the outside world.

They stayed in the main house at the ranch, which had been built in 1857 by a Scottish textile mill owner who grew cotton in Texas for shipping back to his mills. The two-story house was constructed of local limestone with a view of the Hill Country's rocky slopes and oak-dotted plateaus. The Hersheys came only a couple of times a year because Jake didn't care for the ranch as much as his wife. In 1998, they protected the property from future development by negotiating a conservation easement with the Hill Country Land Trust, which Terry and Lady Bird Johnson set up. It is the largest piece of protected land in Gillespie County, and Sansom helped with the transaction.

By the early 2000s, the Sansoms had grown to love the ranch as a second home. It was a reminder of childhood vacations in the Hill Country, where the drier air and cooler water provided a respite from the heat and humidity of the Texas coast. The rivers were clear instead of muddy and the nighttime stars brighter than those dimmed by oil refinery lights. "We dreamed about it when we weren't there," Sansom said. They started thinking about renovating the 150-year-old house, and Sansom hired a researcher to gather documents for the reconstruction, filling ten filing boxes and sparing no expense.

In 2003, however, disaster struck. The house burned down mysteriously, though the Sansoms suspected faulty electrical wiring. Even though they were careful to turn everything off before leaving after the weekend, an electrical short seemed the likeliest explanation. The fire destroyed the house, all of Sansom's research, his library, and his dreams. He and Nona were so devastated by the destruction that afterward they rarely visited. The ruins lay abandoned for a decade, with hackberry trees engulfing the carcass.

While the Sansoms were heartbroken about the disaster, they remained emotionally attached to the ranch. Hershey wanted to give the property to Sansom, though he rejected it as an outright gift and proposed another option. They agreed on a life estate in 2011 in which Sansom acquired the legal right to use the ranch during his lifetime. Thereafter, the property would revert back to the Jacob and Terese Hershey Foundation.

With the life estate in place, Andy and Nona started contemplating a restoration of the house, which was little more than charred stone walls overgrown with brush. They had been reluctant to take on the project because they were told repeatedly that the burned remnants couldn't be brought back to life. However, the Sansoms had connections in the construction industry after years of involvement in the restoration and maintenance of Parks and Wildlife structures. For example, Sansom had selected architect Emily Little to be the project manager for the state cemetery restoration a few years earlier, and she had later remodeled a couple of his homes. Little was known for her renovation of historic buildings, with a focus on preservation of trees and natural terrain. She was an obvious hire for the homestead, given her niche and network of specialized craftsmen. Little found an expert in the rehabilitation of old stone buildings, an engineer from Georgetown who preserved historic dance halls in Texas. He knew exactly what to do.

The contractor for the rehabilitation project was Paul Cimino, a friend of the Sansoms who also had worked on their house in Austin. Cimino also was an artist and had a crew of carpenters and stonemasons who were musicians as well. They played with handmade banjos and guitars at Austin venues on a regular basis. "All of these guys were creative people," Sansom said. "They were artists and musicians and saw this as a labor of love. This was Paul's masterpiece, Nona said, and it was the last thing he did before he died."

Shortly before the reconstruction started, Cimino was told that his congenital lung disease was worsening. It had killed several family members and "was a death sentence," Sansom said. Cimino had told him, "If we're going to do this, we need to get on it." His words "tipped the balance" for the Sansoms, who had been dithering. Halfway through the job, Cimino got sicker and started using bottled oxygen on the jobsite.

Soon thereafter, one of his lungs collapsed, and he was put on life support in hopes of getting a lung transplant. In fact, Cimino was flown by helicopter to San Antonio after organs became available, and he received a double lung transplant. Ten days later, he was back on the job and completed the work in 2013.

Apart from the homestead, another structure also was restored. A small, wooden-frame house had been badly damaged by a hailstorm, leaving broken windows and holes in one wall. The board-and-batten cottage was repaired and provided guest quarters for visitors to the ranch. In 2020, it became the lodging for April, who moved in to be near her new job as executive director of the Bamberger Ranch Preserve.

A community conservationist and wildlife ecologist with a doctorate degree, April had grown up in a household that cherished nature and recognized the responsibility to protect it. After getting her undergraduate degree in wildlife ecology, she worked as a volunteer with the Peace Corps in the Philippines, focusing on community conservation. She seemed ideally suited to run the Bamberger Preserve and carry out its mission of land stewardship, outreach, education, and research. "You can't separate my desire to work in the environmental and conservation fields from my experience growing up," she acknowledged.

After the homestead renovation was done, Sansom turned his attention to the land. He followed the playbook written by his friend, J. David Bamberger, who had pioneered ranch restoration in Texas decades earlier. Native grasses were at the heart of Bamberger's approach, providing the foundation for all other restorative practices.

"Grass is Mother Nature's greatest conservation tool," Bamberger would tell Sansom's students at Texas State University. Like a sponge, grass-covered soil absorbed and held water. "If you have grass on the ground, it's the best conservation thing going," Bamberger told the Conservation History Association of Texas. "You can build diversion ditches and terraces on the landscape. You can work on the creek banks and try to concrete them or rock them in—that will cost you millions if not billions of dollars. There's nothing as quick, nothing as simple, nothing as easy as restoring native grass."

Bamberger once made a legendary appearance at the Texas capitol, where he extolled the virtues of grass in a committee hearing. Recounted in Sansom's book *Seasons of Selah*, the story goes that lawmakers were aghast at a prop that Bamberger pulled out of a garbage bag during his testimony—a big clump of grass with its long roots dangling and dirt dribbling everywhere. Moreover, he continued to talk longer than his allotted time despite being gaveled by the committee chair.

Nearly one hundred species of grasses now carpet the hills of Bamberger's Selah Ranch—keeping soil from washing away, holding water in the ground, and providing a healthy home for animals, including cattle. When Bamberger bought the Hill County ranch in 1969, forage was sparse because the land had been overgrazed and Ashe juniper was rampant. More than forty acres of land were needed for each cow to survive. Over the subsequent forty years, Bamberger removed woody plants such as greenbriers and grapevines to allow indigenous grasses to return. By 2010, only half as many acres were needed for each beeve.

Sansom's ranch suffered from many of the same ills as Bamberger's in the adjacent county, so he started as his role model did by removing invasive plants to help restore habitat health and diversity. Ashe juniper, often referred to as cedar in the Hill Country, was cleared on about one thousand acres, and another four hundred acres were burned to manage brush encroachment.

The next step was getting rid of Kleingrass—an exotic plant imported from South Africa that was grown to feed cattle—so that indigenous grasses and wildflowers could return. The process began with haying the Kleingrass to cart off as much of it as possible in the autumn and then letting it rest before applying two rounds of herbicide over winter. By spring, native plants started to appear.

The second phase of ranch restoration was healing degraded land. Soil erosion was pronounced in some areas, especially ravines where animals clambered down to Williams Creek to drink. Nothing was growing on them anymore, and when it rained, the gullies got slick and muddy, sending excessive amounts of dirt into the creek. "Some of the areas are so eroded, they're almost gorges," Sansom explained. "They're bald as a head."

The Sansoms took the first steps to stop the runoff by trying to force animals off the eroded gullies and onto other paths to the creek. Straw netting was laid at gulley entrances with the intent of scaring away cattle and deer, in particular. The netting, which was similar to those placed along highways for planting grass, did nothing to ward off thirsty animals.

"We laid a number of one hundred-foot rolls of netting, but as soon as we had gotten them laid out, cattle approached and walked right over the netting to get down to the water," Sansom said. He then worked with volunteers to pile up cedar branches to repel cattle and deer, which don't like the prickliness of the foliage. Cutting the cedar branches was more work than the Sansoms could manage because it had to be done within a short time to comply with federal regulations governing the golden-cheeked warbler, an endangered songbird whose habitat can't be destroyed at certain times. Volunteers were recruited from youth hunts held on the ranch.

By this time, Sansom had arranged with Texas Parks and Wildlife to use his ranch for a pilot project to test erosion techniques pioneered by riparian restoration expert Bill Zeedyk and endorsed by the Natural Resources Conservation Service. Zeedyk was sometimes called the "Stream Whisperer" and developed his simple restoration techniques following a career of more than thirty years with the US Forest Service. His low-cost methods used sticks, rocks, and soil to slow and spread water runoff, as described in his book, *Let the Water Do the Work: Induced Meandering, an Evolving Method for Restoring Incised Channels*. Parks and Wildlife wanted to test whether the practices, approved by NRCS in Colorado and New Mexico, would work in Texas, and the agency recommended using a restoration ecologist by the name of Mollie Walton to help with the pilot.

Walton was an acolyte of Zeedyk and worked with nonprofit groups in New Mexico, where she was based, and in Texas, where her family had a ranch in Lampasas. "Texas is pretty cynical," Walton noted. She explained that a rancher at a Texas erosion workshop was asked whether he thought the event was worthwhile. "His answer was, 'I don't know, ask me in three years.'"

The Hershey Ranch project was designed to do two things—repair eroded soil and restore the water cycle. Walton brought a precision to the undertaking that hadn't existed before, using surveying equipment to measure the slope of the terrain and identify where brush should be placed to control water runoff into Williams Creek. The brush was piled into tentlike structures called "pods" that not only controlled runoff, but also served as nurseries for new plants. Birds roosted in the pods, dropping seeds that gave rise to fledgling plants protected from deer and cattle by the surrounding cedar.

In addition, the pods captured sediment so that excessive amounts didn't run into Williams Creek, which fed the Pedernales River, and further endanger the Pedernales River springs salamander. In 2021, the salamander was the subject of a petition to list it as threatened or endangered. In 2022, the US Fish and Wildlife Service said that listing might be "warranted" and added the salamander to its listing workplan.

Sansom's efforts to return the land to a more sustainable state included raising cattle. Cattle eat grass, which allows space for flowers (or forbs) that deer eat, contributing to biodiversity. Sansom leased some of the ranch property to a carefully selected cattleman who was recommended by a top range scientist in Texas. The cattleman was a "modern guy" who used a drone to monitor the herd's movements so that pastures weren't overgrazed.

The animals were regularly moved through various pastures to avoid overgrazing any one of them. To streamline the movements, Sansom constructed a system of fences to facilitate planned rotation around the ranch. Internal fences that had deteriorated were rehabilitated, and electric fences were built for short term, intense grazing intended to mimic bison behavior. Sansom knew something about bison, having coauthored the book *Southern Plains Bison: Resurrection of the Lost Texas Herd*.

"Bison would come in and hit it hard, and then move on," Sansom explained. "The only way you can do that with cows is to move them into a confined area with electric fences. They hit it hard, and then you move the electric fence and move them over."

In 2018, another threat to the ranch emerged, one that seemed to fly in the face of the environmental protection that Terry Hershey had put in place with the creation of a conservation easement. The Hershey Foundation, legal owner of the ranch, received notice of plans to build a natural gas pipeline across the property in a letter from Kinder Morgan, one of the largest energy infrastructure companies in the nation. The company routed the pipeline across the Hershey Ranch and others in the Hill Country as part of its path from the Permian Basin to the Texas coast. Kinder Morgan intended to use eminent domain to condemn a one hundred-foot-wide easement across fourteen acres during construction and fifty-foot-wide thereafter.

Sansom and other landowners argued that the condemnation process overseen by the Railroad Commission of Texas, which permits oil and gas pipelines, was too lax. "The Railroad Commission doesn't require anything except the applicant's word that the pipeline will be a common carrier," Sansom said. "This is outrageous and totally different from the process by which electrical transmission lines and other infrastructure apply for eminent domain."

The landowners sued the Railroad Commission on grounds that due process was lacking and should have included public hearings, delineation of alternatives, an environmental impact statement, and opportunity for protests. The property owners were joined by Hays County and the City of Kyle. Their arguments got some sympathy from the judge.

"The court is concerned with a power that, when exercised by a governmental entity, must be done in the harsh light of public scrutiny of open meetings and public notice, but, when exercised by a private entity, may be determined without public notice by a select few driven primarily by their financial interests," wrote Judge Lora Livingston of the 261st District Court in Travis County. "However, the court must also be conscious of its role to apply the law and not to dictate the policy of the state." Judge Livingston dismissed the suit.

The Permian Highway Pipeline cleared and carved a swath through Hershey Ranch and went into service in December of 2023.

The landowner lawsuit underscored the perception that land protection, restoration, and conservation amounted to a luxury that only the wealthy could afford because of the high costs. The time and money spent by Sansom and his role models—Bamberger and Bromfield among others—looked like a rich man's hobby. While precise costs per acre are hard to come by, landowners needed to have financial resources for expertise and for labor, as well as the knowledge and time to discover and navigate government programs.

Bamberger was wealthy when he began restoring his ranch that became a showcase of environmental renewal in Texas. He told Sansom in a magazine interview that "money no longer meant anything" to him when he was embarking on the endeavor at the age of forty. Thus, he could spend millions of dollars on breathing new life into exhausted land.

While Sansom may not have had as much money as Bamberger, he was financially comfortable and—most importantly—he knew how to navigate government programs that provide funding and resources for his projects. He understood how to apply for programs, implement them, and adhere to timelines and stipulations. Sansom was familiar with wildlife management and habitat management plans—how they were developed and put in place. He knew that the ranch needed a wildlife manager, so he arranged for the Texas Audubon Society's director of conservation strategy to serve in that capacity. Sansom had the financial reserves to maintain and preserve the investments after they were made.

The long-term nature of the work amounted to what Sansom and others have called "cathedral building." He and Nona might not live to see the full result of their work. Perhaps the greatest reward for Sansom was that the effort involved in making the ranch sustainable provided a sense of being grounded. Manual labor connected Sansom to the land in a way that buying and selling property didn't. He was tilling soil, planting today, and reaping tomorrow—not flying over vast tracts of property from the comfort of a private plane.

While Sansom did everything he could to return the land to health, he couldn't control the climate. It was getting hotter and drier.

Chapter 20

Tipping Point

For the first time in Sansom's life, in 2022, he worried about the world he was leaving to his grandchildren. Climate change had reached a tipping point by some measures. Extreme heatwaves and floods caused by global warming were deemed irreversible by the world's preeminent team of climate researchers, the United Nations' Intergovernmental Panel on Climate Change. In the absence of action, rising deaths from heat, increasing species extinction, and coastal inundation no longer would be reversible by 2030, according to the panel. It was thought that as many as half of all species around the world could go extinct by 2050 because of a warming planet.

The acceleration of climate destruction intensified debate over the "Anthropocene," a proposed epoch in Earth history when human activity started to significantly impact climate and ecosystems. Though a panel of geologists formally rejected declaring the start of a new epoch of geologic time in March 2024, the Anthropocene had been discussed by scientists and widely used in mainstream media. Sansom viewed the concept as a "a complicated narrative," and "the story of creation in the sense that the central element is what we are doing to the planet."

In Texas, climate change was more visible than in most states. Excessive heat, extreme drought, and wildfires hit Texas more often than all but a handful of other states. During the first quarter of the twenty-first century, Texas suffered its worst one-year drought in history in 2011, when several cities came within days of running out of water. The state could expect twice as many one hundred-degree days and more intense drought over the next fifteen years, according to a 2021 study by the state climatologist, Texas A&M University, and the Texas 2036 public policy think tank.

If the state has another drought of record—the worst in its history—water supplies would run short by more than 25 percent, based on the 2022 State Water Plan of Texas. The drought of record was 1950–57, the most severe dry spell that had occurred in Texas up to that time. That drought became the legal test of whether water supplies are "firm" for contractual purposes, meaning they must be available during a drought as bad as the 1950s. Sansom and most scientists believe that worse droughts are likely to come because worse ones have occurred in the past.

In 2022, a majority of Texans wanted their elected officials to do something about climate change, according to Yale's Program on Climate Change Communication and George Mason University's Center for Climate Change Communication. Yet many local and state officials in Texas remained silent on global warming, denied it, or argued that combatting it would slow economic growth. While the national Republican Party had softened its stance on climate change by the 2020s, the Texas GOP lagged behind. In 2022, Republicans controlled every statewide office in Texas, both houses of the legislature, the entire state supreme court, and both US Senate seats. Elected officials felt little urgency to respond to constituent calls for climate action because of insulation from voter pressure provided by electoral gerrymandering and strict voting laws.

The Texas GOP's intransigence put it out of step with much of the state's business community, which was actively devising strategies to adapt to global warming. A growing number of oil and gas, engineering, manufacturing, and service companies were laying out transition paths away from fossil fuels and toward clean energy—even in petroleum-dependent Texas. In Houston, the energy capital of the world, business leaders were capitalizing on the city's expertise to develop low-carbon technologies in a project called the Houston Energy Transition Initiative.

Sansom, a lifelong Republican, was disheartened by his party's skepticism about settled science on climate change and its embrace of extreme ideologies—a contrast with the party's pragmatism and science-based policies of the past. In 1980, he attended the Republican National Convention as a delegate for George H. W. Bush, though he ultimately voted for Ronald Reagan, who chose Bush as his running mate. "If you look back at many of the things that Reagan said, they look progressive compared to the Republican leadership of the present day," Sansom said.

About 97 percent of climate scientists agree that humans are causing global warming, according to the American Association for the Advancement of Science. Yet despite the preponderance of evidence linking climate change to human activity, most state agencies in Texas avoid use of the term "climate change" or "global warming" in formal documents, and Governor Greg Abbott has said he doesn't know whether the earth is warming because he's not a scientist.

One of Sansom's grandchildren echoed his grandfather's concern. "I, sadly, do not see the ability nor the willingness of the people nor the governments of the world to create the massive and rapid changes needed to truly lessen the impacts," said Alex Sansom in an email squeezed in between his studies at Rice University. "Work like my grandfather's—small, concentrated, and passionate work to maintain the natural earth where we still can—seems to be the only work having any actual positive impact."

Given the tepid response from government, ad hoc efforts are gathering pace, according to the young Sansom. "One of my grandfather's students and friends described two camps: 'on-the-ground' versus 'in-the-clouds' conservation," he explained. "The 'in-the-clouds' conservation consisting of scientific research, computer models, and government policies is very much separate from the 'on-the-ground' work of land stewards like my grandfather and my aunt, April Sansom. These two camps of climate work must converge to build a better future. I have hope for that, but I am most definitely concerned."

Still, the elder Sansom remained drawn to the GOP, even as he acknowledged inner conflict. "I believe the current direction of the Republican Party places in danger the future of democracy in our country," he said. "I am grateful to state political leaders like Joe Straus who have stood up against the tyranny of many Republican politicians today, and I am a very strong supporter of Liz Cheney (although not all of her policy positions) and have known her since she was three years old."

Straus and Sansom had long been members of the "WannaMeetaGOP" caucus, which formed in the 1970s to revive the Republican Party in Texas by promoting low taxes, small government, and limited services.

Sansom rued the political calcification that undermined both the Republican and Democratic parties in the 2020s. Electoral gerrymandering enabled polarization and left a void in the middle, where he always had felt most comfortable. In fact, Sansom believed that elected officials and party bosses in both parties had moved to more extreme positions than their constituents.

An old-fashioned Republican, Sansom was likened to Teddy Roosevelt by some. "Andy is the embodiment of Teddy Roosevelt," said Ken Kramer, a doyen of environmentalism in Texas. Sansom protected more than half-a-million acres of land in more than one hundred parks, wildlife management areas, and other protected lands in the same way that Roosevelt protected over 230 million acres in 150 national forests, fifty-one federal bird reserves, four national game preserves, five national parks, and eighteen national monuments.

Some twenty-five years after Teddy Roosevelt was president, his distant cousin Franklin D. Roosevelt provided another inspiration for Sansom. FDR's New Deal was the kind of policy response demanded by the climate crisis, in Sansom's opinion. The New Deal's Civilian Conservation Corps provided a model for the Civilian Climate Corps proposed by President Joe Biden in his landmark 2022 climate change legislation. The Biden CCC would provide jobs in national parks and remote areas to create trails and restore wetlands, build green infrastructure to reduce stormwater runoff and flooding, and retrofit homes to be more energy efficient.

The longer-term view of natural resource protection taken by the Democratic Biden administration aligned more with Sansom's philosophy than the shorter-term view of the previous Republican administration, led by President Donald Trump. The Trump administration favored exploitation of natural resources over conservation of them. For example, Trump opened the Arctic National Wildlife Refuge, one of the most pristine ecosystems in the world, to oil drilling, while subsequently Biden froze the leases, pending review.

In Texas, the shorter-term view often had been invoked on the grounds that future technology would come to the rescue before natural resources were completely exhausted. Hence, groundwater was being monetized at an accelerating pace in the first decades of the twenty-first century. The

Ogallala Aquifer, the largest in the country, underlies much of the Texas High Plains and was being mined—meaning more water was being taken out through pumping than nature was putting back in through recharge.

In the absence of clear governmental leadership to address the climate crisis and environmental degradation, grassroots efforts gathered pace in the United States. Nearly 14,600 nonprofits worked to protect the environment and animals in 2015, according to the Urban Institute's National Center for Charitable Statistics. The bottom-up approach to environmental protection was inefficient because of redundant efforts, duplicated overhead, lack of resources, inexperience, and competition for donations, according to various studies. Still, the fragmented approach held enormous attraction in a state like Texas, where small government and limited services were preferred.

Sansom had much to offer nonprofits focused on the environment—a prodigious ability to fundraise for charity, a career devoted to the protection of nature, countless personal relationships across the state, an ability to find common ground, and an intuition about which way the wind was blowing before acting. In 2022, he sat on the boards of directors of Bat Conservation International, the Texas Water Foundation, and Friends of Blue Hole Park, which he helped found.

In addition, he was on the board of the Jacob and Terese Hershey Foundation, which funded an initiative close to his heart—the Great Springs Project. It was aimed at connecting four of Texas' most beloved springs to create a greenway between Austin and San Antonio. The greenway was intended to protect lands over the Edwards Aquifer recharge zone while linking trails between Barton, San Marcos, Comal, and San Antonio springs.

Sansom also served on the steering committee of the Million Acre Parks Project, an effort spearheaded by the nonprofit Environment Texas to advocate for $1 billion in state funds to buy one million acres of new parkland by 2030. The Parks Project steering committee included actor Ethan Hawke, a native Texan who's been nominated for four Academy Awards. Hawke had worked with Environment Texas before, having been recruited by the group's indefatigable executive director, Luke Metzger. The Million Acre Parks Project was particularly meaningful for Sansom, who had doubled the amount of state parkland while helming

TPWD. Yet in 2022, Texas ranked only thirty-fifth in the country in parkland per capita, according to the National Association of State Park Directors.

Against all odds, the steering committee succeeded in persuading the 2023 Texas legislature to appropriate the money, a remarkable feat considering lawmakers' historical reluctance to fund even the maintenance of state parks. The $1 billion appropriation required voter approval, which it won a few months later, amounting to "the biggest investment in nature in Texas history," according to Metzger.

By his own estimate and others, Sansom raised more than $240 million for various environmental causes that he created or led over his lifetime—$20 million for The Nature Conservancy, $200 million for the Texas Parks and Wildlife Foundation, $20 million for the Meadows Center, and $500,000 for Texas A&M University Press.

While dismayed by climate change and political polarization, Sansom nevertheless continued his lifelong work to protect the environment. He did not rest on his laurels, which included at least a dozen professional awards ranging from the Lifetime Achievement Award given by The Nature Conservancy to the Chuck Yeager Award given by the National Fish and Wildlife Foundation. In addition, the National Parks Foundation awarded him Cornelius Amory Pugsley Medal, which recognizes outstanding contributions to the promotion and development of public parks and conservation in the United States.

In 2020, Sansom was inducted into the Texas Institute of Letters, an august honor society founded in 1936 to celebrate the state's literature and recognize distinctive literary achievement. He was the author of nine books, general editor of twenty-three books—including several series—and the writer of numerous magazine articles.

After leaving the executive directorship of the Meadows Center in 2020, Sansom continued his association with the center as its founder and as a professor in the university's department of geography. In addition, he did his own consulting on a variety of projects that sought to ensure sustainability of natural resources.

One of his projects involved the iconic Comanche Springs in West Texas, once an oasis in the northern part of the Chihuahuan Desert that provided water for Native Americans, explorers, and farmers starting in the pre-Columbian era. In 1936, the New Deal's Works

Progress Administration constructed a magnificent bathhouse at the site of the springs in Fort Stockton, and the city launched its historic Water Carnival.

In 1950, swimmers in the pool noticed that bubbles no longer were floating up from the bottom, prompting the city and farmers to investigate. Their experts said the springs had gone dry because aquifer levels had fallen due to heavy groundwater pumping from irrigation wells. After negotiations with well owners failed, Pecos County filed a lawsuit in a bid to regulate pumping for protection of the springs. In 1954, the court ruled that the groundwater feeding Comanche Springs belonged to the landowners, who could do as they pleased, even if it dried up the springs used by the community. The families who relied on the springs to irrigate their crops were devastated, and by 1961, the springs quit flowing.

In the early 1990s, the springs trickled out again, and by 2022 they flowed for a few weeks a year. A few farmers were left in Pecos County, including descendants of Clayton Williams, one of the landowners named in the 1954 suit. Williams, who ran unsuccessfully for governor of Texas in 1990, was asked at the time by *Texas Monthly* whether he thought Comanche Springs could flow again if he stopped pumping thirty million gallons of groundwater daily. "They might," he answered. "But I'm not going to do it. It's my land, and I have the right to use the water. . . . I'm a businessman. I'm a cow man. I'm a conservationist. I didn't dry up those springs. I bought the land. It's mine, and if I didn't pump water, it wouldn't be worth anything."

Williams' insistence that he could pump all the water he wanted was based on the rule of capture doctrine that underpins groundwater law in Texas. The doctrine even allows landowners to siphon off groundwater from their neighbors, unless constrained by a groundwater conservation district or a lawsuit. In 2009, a Williams family holding company applied for a permit from the Middle Pecos Groundwater Conservation District to export groundwater to Midland-Odessa, prompting concerns about the potential impacts on local groundwater users and the Williams' ability to monetize their neighbors' assets.

"Many challenges still exist for effective and sustainable management of groundwater in the future, including antiquated state laws and rules, the potential for vulnerable districts being controlled by parties interested only in export, areas that refuse to create groundwater districts, and the ever-present demand of large municipalities," Sansom wrote in his book *Water in Texas.*

In 2017, a modified groundwater export permit sought by the Williams family's Fort Stockton Holdings Company was granted, causing some to worry about whether the groundwater planning process was working adequately to protect the aquifer. One of those most worried about rigor in the permitting and planning process was Ernest H. Cockrell, an oil and gas executive whose family owned one of the remaining farms in Pecos County that relied on groundwater for irrigation.

Cockrell's father, an oilman and who endowed the Cockrell School of Engineering at the University of Texas, had established a pecan farm near Fort Stockton in the 1960s. Belding Farms was named after the bountiful part of the Edwards-Trinity Aquifer known as the Leon-Belding area. The younger Cockrell and his wife, Janet, had lived in Fort Stockton in the early years of their marriage, and he remained actively engaged in the farming operation thereafter.

In 2017, a Cockrell family company that owned the farm sued the groundwater district over the export permit granted to Fort Stockton Holdings due to concerns about impacts on long-term sustainability of the aquifer. Cockrell needed reliability of the water for his pecan farm, among the largest in the state, and the withering of Comanche Springs wasn't reassuring.

"Comanche Springs was the canary in the coal mine for our community," said Cockrell. "Just as canaries in coal mines warned of threats to human health, the springs warned us of threats to environmental health. In fact, springs are one of the best gauges of the health of most major aquifers in Texas."

Cockrell didn't want to see Edwards-Trinity Aquifer permanently over-pumped, and in 2020, he retained Sansom to help navigate the regulatory and legislative shoals to that end.

A fundamental challenge to the sustainability of the aquifer was the inherent conflict between groundwater as a private property right and as a natural resource—a tension made worse by rule of capture. "What's mine is mine and what's yours is mine," is the phrase sometimes used to describe the rule.

Sansom opposed the "law of the biggest pump" as a way of monetizing neighbors' groundwater, though he believed that private markets offered the best protection against that. As he explained to Votteler of the *Talk+Water* podcast, Sansom said that market-based tools used by The Nature Conservancy could be used to protect water.

"My assessment is that there are two major impediments" to efficient water allocation, Sansom told Votteler. The first one was a lack of pumping caps on groundwater wells, which made water markets unfeasible. "Any landowner who wants to sell his water can move over a mile and drill another well," thereby increasing supply and skewing the pricing mechanism.

The second impediment applied to surface water, where market mechanisms fell victim to the prior appropriation doctrine that enables holders of older water rights to get their water before holders of newer rights. "You could purchase a significant amount of surface water, but there may be a senior holder downstream," who makes a withdrawal and leaves you with nothing.

The dual nature of groundwater—private property and natural resource at the same time—accelerated competition for it as the population of Texas exploded. Growing cities increasingly were looking for water in rural areas where landowners were willing to sell it. One of the most prominent examples in Sansom's lifetime was the Vista Ridge project in which the City of San Antonio bought groundwater that was more than 140 miles away in Burleson and Milam counties. The San Antonio Water System, the city's water utility, paid an estimated $3.4 billion to build a pipeline across Central Texas and convey about fifty thousand acre-feet of groundwater a year from the Carrizo-Wilcox Aquifer.

Impacts to the aquifer had been significant, Sansom said. "Those people are going up to the legislature every day saying, 'I'm losing my groundwater. I've had to lower my well three times in the last five years,'" Sansom noted. "I think that that is going to end up being an epiphany for legislative leaders."

Water needed to be moved around the state because the sources are sometimes in different places than where the demands are, Sansom conceded. Still, he was acutely aware of the risks involved in water transfers, such as protecting water rights and ensuring sufficient future supplies in areas exporting water. "If you establish a combination of thresholds and a marketplace, then you could probably do it," he said. Groundwater pumping thresholds typically provide triggers that entail some action, such as cutbacks in production. Mitigation funds for well owners also can be tapped.

Water markets were required, in Sansom's opinion. "For too long, we have treated water as if it were free and not a commodity with value like petroleum or corn," Sansom wrote in *Water in Texas*. "A major component of conserving our water resources in Texas will be market-based acquisitions of surface water and groundwater rights." These rights need to be traded in a marketplace where players can invest and expect a profit, he added.

Another challenge was the view that water for people and water for nature were competing interests. The competition view was particularly deep-rooted in dry regions where water historically was scarce, such as West Texas. In Pecos County, where Comanche Springs flows only a few weeks a year, even the ephemeral flow through old irrigation canals was seen as a waste by some.

"We need to figure out how to use that water in the canal when the springs flow," argued Pecos County Judge Joe Shuster during a county commissioners meeting in 2022. "It doesn't do any good for that water to flow down the canal."

Sansom argued to the contrary. Environmental integrity was directly tied to water security. "If we do not address the serious environmental issues associated with our growing demand for water, the development of new water resources itself will be jeopardized," he noted in *Water in Texas*. When healthy rivers and aquifers provide water for cities, businesses, and farms, they also sustain plants and animals. The human and wildlife ecosystems are mutually dependent.

Sansom had a bent for operating effectively in overlapping spheres, a Venn diagram of personal and professional projects. His enthusiasm occasionally was greater than his ability to execute because his attention was spread too thin. "Andy sometimes overcommitted and underdelivered," observed one colleague. Still, a cultivated skill was his judgment in pushing enough to achieve a goal, but not so much that relationships were ruptured.

"Andy is a renaissance man," said Kramer. "He sees issues from a larger context. He looks beyond policy to other parts of our society."

One of Sansom's intrinsic talents was his ability to connect with people and extend kindness, which was particularly effective when

accompanied by his trademark chortle. A notable act of kindness
was to befriend a Vietnam veteran by the name of Brandon Lairson.
Around 2007, Lairson walked into a Wednesday night potluck dinner
at Central Presbyterian Church in Austin. Unemployed and homeless,
he was invited to join the meal.

The next thing Sansom knew, Lairson was in the kitchen washing
dishes. For the following eleven months, the congregational pastor allowed him to live secretly in the church. Eventually, Lairson became a
member of the congregation and then a deacon, assisting with the day-
to-day work of the church, such as maintaining the church building and
grounds.

While Lairson didn't have a lot of material possessions, he shared
everything he had. He met a woman who became his life partner, though
she was addicted to heroin. He supported her emotionally and financially
in an eleven-year journey to sobriety. When they married, Sansom
served as Lairson's best man.

Sansom and Lairson maintained a deep friendship, sometimes talking
three times a day. "Brandon is one of my very closest friends from whom
I have received much more than I have given over the years I have known
him," Sansom said.

Sansom was pragmatic and flexible, always looking for practical
solutions to problems in ways that all parties could live with. "Andy
was able to nimbly navigate," explained John Howard, who worked
with Sansom as Governor Bush's environmental adviser. When the
Endangered Species Act provoked backlash from Texas landowners in
the 1990s, Sansom responded with conservation easements as a way to
defuse political ire with financial incentives. "If he was just going to fight
for the Endangered Species Act, he would never have gotten anything
done," Howard said.

Sansom's willingness and ability to find common ground pointed to
a defining trait—balance. His ego was tempered by humility. His role as
an insider occasionally was punctuated by flights of fancy as an outsider.
He stayed grounded and "remained constant," noted Kramer. Sansom's
goal in life was to keep things in a steady state so they did not fall apart.
He was ambitious, yet not aggressive. He saved countless plants and
animals through imagination and resolve.

"He had a huge, crazy vision," Howard said of Sansom's tenure at Texas Parks and Wildlife. "Most people view the executive directorship of an agency as a caretaker role. He blew that up. He said that's not nearly enough. 'I'm going to do way more than anybody ever imagined for this role.'"

A critical part of his vision was the ability to create an emotional connection between humans and their environment. Sansom tapped into an ancient affinity that has existed for millennia but weakened amid rampant urbanization and resource exploitation. "A majority of people feel that land conservation is important," noted Kramer. "Andy was able to show it's important to everyone."

Sansom's bequest to future generations was the preservation of habitat for plants and animals that have little defense against the ravages of *Homo sapiens*. "Andy and land acquisition are synonymous to me," said Jim Blackburn, one of the foremost environmental lawyers in Texas. "His huge achievement is that he was one of the early leaders in the private-sector acquisition of nature reserves on a large scale. And then at Parks and Wildlife, he took and just exploded that. And that to me is a fabulous legacy."

Sansom followed in the footsteps of environmental conservationists such as Stewart Udall, who established four national parks, six national monuments, eight national seashores and lakeshores, nine national recreation areas, twenty national historic sites, and fifty-six national wildlife refuges while he ran the Interior Department. As interior secretary, he recruited minorities for employment, including the first African Americans as park rangers. Udall also warned of environmental pollution, natural resource depletion, and dwindling open spaces in his 1963 book, *The Quiet Crisis*.

Another pioneer of conservation was John Muir, who helped to establish Yosemite, Sequoia, Mount Rainier, and Grand Canyon national parks. A racist, Muir saw these public lands as primarily for white visitors, whereas Sansom sought to make nature more accessible to a wider audience—people of color, families, and outdoor enthusiasts beyond hunting and fishing.

Some of Sansom's legacy was left to wither on the vine, unfortunately. The water resources program he created at Texas Parks and Wildlife to protect rivers, streams, and bays "was allowed to waste away," noted Dan Opdyke, an environmental engineer who formerly worked in the program. Staffing shrank to a fraction of its previous size, and lack of regulatory authority undermined efforts. "When fighting for water for fish, managers knew that we were playing poker with never more than a pair of deuces," Opdyke said. "How many times can you lose before you move on to other priorities?"

The extensive research on environmental flows done by Sansom and his colleagues at Texas State University languished as the e-flow regulatory process overseen by the Texas Commission on Environmental Quality was all but abandoned.

Despite professional—and personal—disappointments, Sansom remained an innate optimist throughout his life. He saw opportunity where others saw obstacles and was willing to persevere when others wanted to quit. "That's my way," he said, with a simplicity that explained much.

Reflecting on his life as he neared his eightieth birthday, he saw his family as his greatest accomplishment. "I share with my wife, Nona, the fact that our children are healthy, educated, compassionate, and moral and that one of them is raising two spectacular grandchildren while the other is pursuing a career in conservation." He regretted not spending more time with his children when they were young, though he sought to make up for it later by arranging one-on-one trips with them to outdoor destinations.

Family, and particularly Nona, provided the counterbalance to Sansom's drive and ambition throughout his life. She was the ballast. "His preoccupation with his many projects and his many different jobs" were the biggest challenges, Nona admitted. Office hours often extended into the night, and business travel sometimes lasted months. "It was hard at times. But I always knew that I really, truly loved and admired him—and that we could get straightened out."

Seeking balance was the essence of Sansom's life—restoring it to the earth and its inhabitants. When bison, prairies, and wetlands suffered from too much humanity, he aimed to restore the ecosystems to sustainability. He sought equipoise between ecology and economics, between public and private interests, and between stewardship and exploitation of nature. Sansom's zodiac sign was Libra, the scales, which symbolized his quest, professionally and personally.

What gave him hope for our planet was the growing realization that we humans must balance our needs with those of nature.

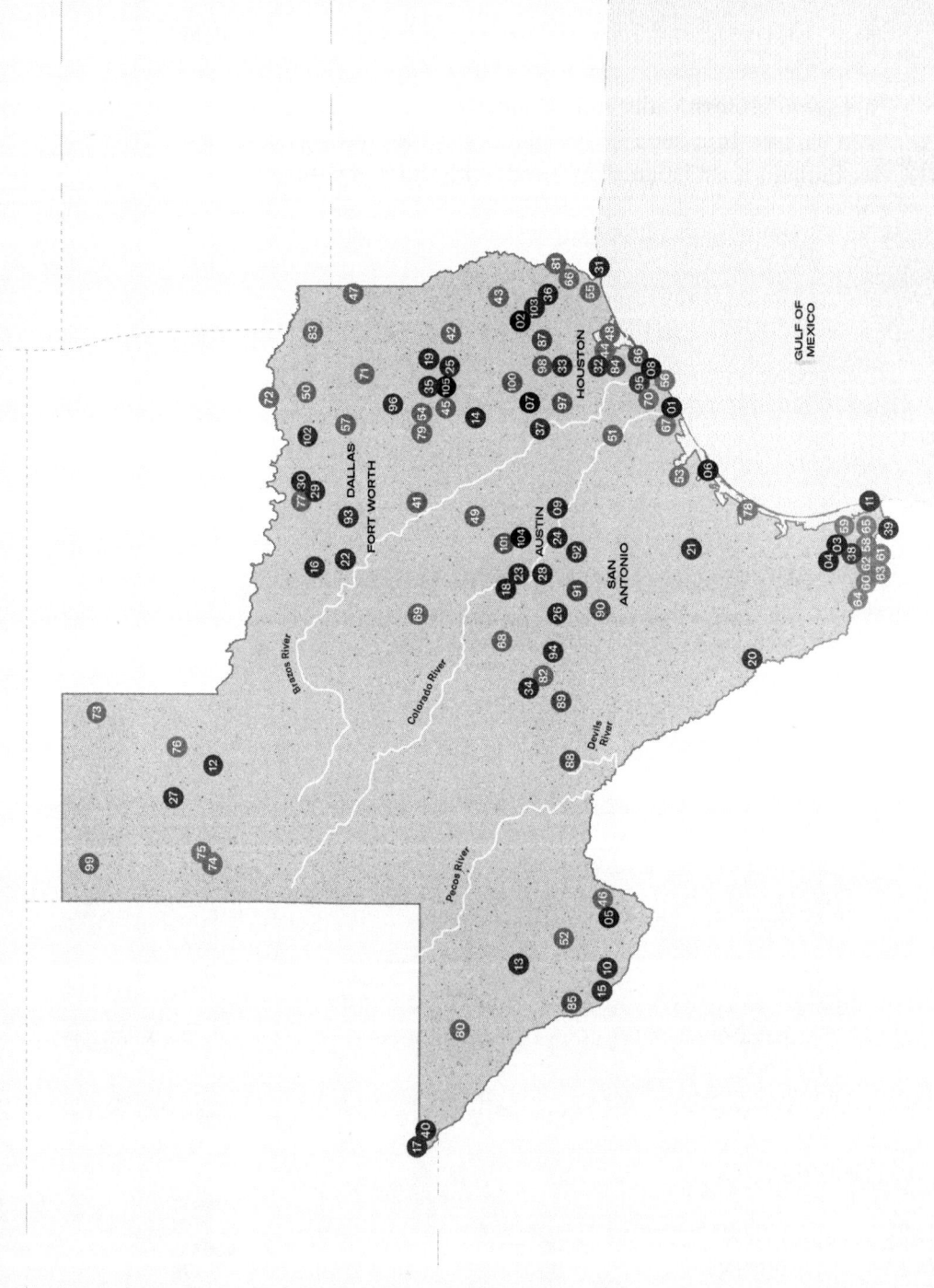

GULF OF
MEXICO

DALLAS
FORT WORTH
HOUSTON
AUSTIN
SAN
ANTONIO

Brazos River
Colorado River
Devils River
Pecos River

Appendix

Critical Lands Protected with Sansom's Help

MAP NUMBER	PROPERTY NAME	CLASSIFICATION	ACREAGE
National Parks, Preserves, and Wildlife Refuges			
1	Big Boggy NWR	National Wildlife Refuge	211
2	Big Thicket National Preserve	National Preserve	113,114
3	Lower Rio Grande Valley NWR, Lower	National Wildlife Refuge	333
4	Rio Grande Valley NWR, Schaleben Tract	National Wildlife Refuge	1,527
5	North Rosillos Mountain Ranch Big Bend National Park	National Park	67,214
6	Matagorda Island NWR and State WMA	National Wildlife Refuge	56,688
7	Sam Houston National Forest	National Preserve	161,000
8	Slop Bowl Tract, Brazoria County	National Wildlife Refuge	1,965
State Parks and Historic Sites			
9	Bastrop SP	State Park	2,272
10	Big Bend Ranch SP	State Park	207,846
11	Boca Chica SP	State Park	1,056
	Caddo Lake SP	State Park	1
12	Caprock Canyons State Park and Trailway	State Park	1,353

	Confederate Reunion Grounds SHS	State Historic Site	3
13	Davis Mountains SP	State Park	1,360
	Eisenhower Birthplace SHS	State Historic Site	3
14	Fort Boggy SP	State Park	1,726
15	Fort Leaton SHS	State Historic Site	6
16	Fort Richardson SP and HS	State Park	24
17	Franklin Mountains SP	State Park	7,858
18	Inks Lake SP	State Park	19
19	Jim Hogg SHS	State Historic Site	177
20	Lake Casa Blanca International SP	State Park	696
21	Lake Corpus Christi SP	State Park	50
22	Lake Mineral Wells State Park and Trailway	State Park	395
23	Longhorn Cavern SP	State Park	25
24	McKinney Falls SP	State Park	72
25	Mission Tejas SP	State Park	227
	National Museum of the Pacific War	State Historic Site	4
26	Old Tunnel SP	State Park	16
	Palmetto SP	State Park	1
27	Palo Duro Canyon SP	State Park	2,138
28	Pedernales Falls SP	State Park	352
	Port Isabel Lighthouse SHS	State Historic Site	0
29	Ray Roberts Lake SP, Greenbelt Trailway	State Park	1,300
30	Ray Roberts Lake SP, Isle Du Bois	State Park	4,238
31	Sabine Pass Battleground SHS	State Historic Site	58
	Sam Bell Maxey House SHS	State Historic Site	0
32	San Jacinto Battleground SHS	State Historic Site	1,200

33	Sheldon Lake State Park and Environmental Learning Center	State Park	215
34	South Llano River SP	State Park	16
	Starr Family SHS	State Historic Site	3
35	Texas State Railroad Loop Park	State Park	25
36	Village Creek SP	State Park	107
37	Washington-on-the-Brazos SHS	State Historic Site	136
38	World Birding Center, Estero Llano Grande SP	State Park	73
39	World Birding Center, Resaca de la Palma SP	State Park	100
40	Wyler Aerial Tramway (Franklin Mountains SP)	State Park	199
State Wildlife Management Areas			
41	Aquilla WMA	State Wildlife Management Area	9,700
42	Alazan Bayou WMA	State Wildlife Management Area	2,074
43	Angelina-Neches/Dam B WMA	State Wildlife Management Area	12,636
44	Atkinson Island WMA	State Wildlife Management Area	152
45	Big Lake Bottom WMA	State Wildlife Management Area	4,540
46	Black Gap WMA	State Wildlife Management Area	4,266
47	Caddo Lake WMA	State Wildlife Management Area	6,070
48	Candy Abshier WMA	State Wildlife Management Area	209
49	Cedar Creek WMA	State Wildlife Management Area	159

50	Cooper WMA	State Wildlife Management Area	9,460
51	D. R. Wintermann WMA	State Wildlife Management Area	246
52	Elephant Mountain WMA	State Wildlife Management Area	23,147
53	Guadalupe Delta WMA	State Wildlife Management Area	2,977
54	Gus Engeling WMA	State Wildlife Management Area	18
55	J.D. Murphree WMA	State Wildlife Management Area	887
56	Justin Hurst (Peach Point) WMA	State Wildlife Management Area	10,311
57	Lake Tawakoni WMA	State Wildlife Management Area	1,562
58	Las Palomas WMA, Anacua Unit	State Wildlife Management Area	22
59	Las Palomas WMA, Arroyo Colorado	State Wildlife Management Area	20
	Las Palomas WMA, Champion Unit	State Wildlife Management Area	2
60	Las Palomas WMA, Chapote Unit	State Wildlife Management Area	220
61	Las Palomas WMA, Ebony Unit	State Wildlife Management Area	266
62	Las Palomas WMA, Kiskadee Unit	State Wildlife Management Area	14
63	Las Palomas WMA, MacWhorter Unit	State Wildlife Management Area	24
64	Las Palomas WMA, Taormina Unit	State Wildlife Management Area	607
65	Las Palomas WMA, Tucker Unit	State Wildlife Management Area	74
66	Lower Neches WMA	State Wildlife Management Area	4,580

67	Mad Island WMA	State Wildlife Management Area	1,560
68	Mason Mountain WMA	State Wildlife Management Area	5,301
69	McGillivray and Leona McKie Muse WMA	State Wildlife Management Area	1,973
70	Nannie M. Stringfellow WMA	State Wildlife Management Area	3,552
71	Old Sabine Bottom WMA	State Wildlife Management Area	5,279
72	Pat Mayse WMA	State Wildlife Management Area	8,925
73	Pat Murphy WMA	State Wildlife Management Area	889
74	Playa Lakes WMA, Armstrong Unit	State Wildlife Management Area	160
75	Playa Lakes WMA, Dimmit Unit	State Wildlife Management Area	420
76	Playa Lakes WMA, Taylor Unit	State Wildlife Management Area	530
77	Ray Roberts Lake WMA	State Wildlife Management Area	40,920
78	Redhead Pond WMA	State Wildlife Management Area	37
79	Richland Creek WMA	State Wildlife Management Area	20
80	Sierra Diablo WMA	State Wildlife Management Area	2,560
81	Tony Houseman WMA	State Wildlife Management Area	3,896
82	Walter Buck WMA	State Wildlife Management Area	16
83	White Oak Creek WMA	State Wildlife Management Area	25,500

Star Natural Areas and Costal Preserves			
84	Armand Bayou CP	State Coastal Preserve	290
85	Chinati Mountains SNA	State Natural Area	37,885
86	Christmas Bay CP	State Coastal Preserve	4,173
87	Davis Hill SNA	State Natural Area	56
88	Devils River SNA	State Natural Area	19,989
89	Devil's Sinkhole SNA	State Natural Area	58
90	Government Canyon SNA	State Natural Area	12,244
91	Honey Creek SNA	State Natural Area	469
Sate Fish Hatcheries and Laboratories			
92	A. E. Wood SFH	State Fish Hatchery	6
	Dickinson Marine Lab	State Marine Lab	4
	Dundee SFH	State Fish Hatchery	1
93	Fort Worth Fisheries Office	State Fish Hatchery	61
94	Heart of the Hills Fisheries Science Center	State Fish Hatchery	55
95	Sea Center Texas	State Fish Hatchery	75
96	Texas Freshwater Fisheries Center	State Fish Hatchery	98
Regional, County, City, and Academic Properties			
97	Kleb Woods Nature Preserve	County Preserve	133
98	Lake Houston Wilderness Park	City Park	4,787
99	Lake Rita Blanca City Park	City Park	1,668
100	Pineywoods Environmental Research Laboratory	Academic Research Laboratory	247
101	Twin Lakes Park	County Park	50

The Nature Conservancy			
102	Clymer Meadow Preserve	Private—The Nature Conservancy	113
103	Roy E. Larsen Sandyland Sanctuary	Private—The Nature Conservancy	5,654
Other Nonprofits			
104	Bright Leaf Preserve SNA	Nature Preserve	217
	Hooks Woods	Nature Preserve	2
105	Ivy Payne Preserve	Nature Preserve	465

Notes on sources: This data is drawn from the Texas Parks and Wildlife Department, The Nature Conservancy, and the websites of relevant properties. The names and sizes of the properties are based on 2023 data, not the data at the time of the properties' establishment. All acreage numbers are rounded to the nearest whole number. Zero acres equates to less than five acres.

—

Notes

The idea for this book was born on the cusp of the COVID-19 pandemic, and by the time I started my research in January 2020, warning lights were blinking around the world. Over the next couple of months, lockdowns, masks, social distancing, supply-chain shortages, and uncertainty about any return to normality ensued. The deadly disease added a layer of drama to the book project, which already was a foray into uncharted waters for me. I'd never written a book before.

I decided to carry on as though everything *was* normal. I interviewed Andy online, which was fine because we knew each other well. Interviews with others proved a little more challenging. When I interviewed in person, we had to figure out masks and where to place the tape recorder so I could still reach it while being six feet away from the interview subject. Fairly quickly, Zoom video calls gained acceptance for all manner of communications.

Less certain was the whole book process, given bottlenecks in paper and other supplies.

By early 2022, the world had relaxed enough that I could visit Andy and Nona at the Hershey Ranch and meet in person with Mollie Walton, the erosion control guru who restore gouged-out ravines.

By 2023, the pandemic had subsided, and the book was completed. What follows is a short outline of sources used, by category.

Chapter 1
Interviews
Kay Bailey Hutchison, Andrew Sansom
Other Sources
Alaska.org, explorenorth.com

Chapter 2
Interviews
Andrew Sansom

Other Sources

Alden B. Dow Home & Studio website, Dow Chemical Company, *History of Lake Jackson* (MacLean), *Scout, the Christmas Dog* (Sansom and Guzman), Texas State University Oral History Project, *Untapped New York*

Chapter 3

Interviews

Andrew Sansom, Nona Sansom

Other Sources

Austin College, Center for Labor Economics, Congressional Research Service, Khan Academy, Lake Texoma, Legacy.com, Lodge and Resort, *Nation*, National Recreation and Park Association, Study.com, University of California, Berkley

Chapter 4

Interviews

Andrew Sansom

Other Sources

Business Wire, Earth Day, History.com, International Institute for Sustainable Development, Resources for the Future, Texas State University Oral History Project

Chapter 5

Interviews

Andrew Sansom, Nona Sansom

Other Sources

Anchorage Daily News, Factsanddetails.com, History.com, Mission: Wolf, *New York Times*, *Realms of Beauty: The Wilderness Areas of East Texas* (Fritz and Alford), RememberSingapore.org, Rogers C. B. Morton Collection, *Star Democrat*, *Texas Monthly*, *Texas Parks & Wildlife Magazine*, Texas State University Oral History Project, UC Berkeley Rausser College of Natural Resources, *Washington Post*, Wolf Song of Alaska

Chapter 6

Interviews

Andrew Sansom

Other Sources

Battlefield: Vietnam (PBS), Calhoun County Museum, Congressional Record, *Handbook of Texas Online*, Island Conservation, Planeta.com, *Texas Lost: Vanishing Heritage* (Sansom, Reid, Meinzer), Texas State University Oral History Project, The Free Dictionary, US Commission on Ocean Policy

Chapter 7

Interviews

Jim Blackburn, Andrew Sansom, Nona Samson

Other Sources

Calming the Waters, Buffalo Bayou: An Echo of Houston's Wilderness Beginnings (Aulbach), CNBC, *George P. Mitchell: Fracking, Sustainability, and an Unorthodox Quest to Save the Planet* (Steffy), *Houston & Nature, Houston Chronicle, New York Times, Public Administration Review, Texas Co-Op Power,* Texas Legacy Project, *Texas Monthly*

Chapter 8

Interviews

Kay Bailey Hutchison, Karl Rove, Andrew Sansom, Nona Sansom

Other Sources

Conservapedia.com, *Handbook of Texas Online, Texas Observer,* Texas State University Oral History Project, Untermeyer.com, Yellowpages.com

Chapter 9

Interviews

Jim Blackburn, Ken Kramer, Andrew Sansom, Andrew Sansom Jr.

Other Sources

Austin Bulldog, Austin Chronicle, Economic History Association, *Encyclopedia Britannica,* History.com, *Journal of Policy History, Los Angeles Times,* National Audubon Society, *New York Times,* Public Citizen, State Bar of Texas, StateImpact/NPR, *Texas Environmental Law Journal,* Texas Legacy Project, Texas Natural Resource Server, *Texas Observer,* Texas Public Policy Foundation, *Texas Tribune, The Texas Water Plan* (Texas Water Development Board)

Chapter 10

Interviews

J. David Bamberger, Ken Kramer, Andrew Sansom

Other Sources

Encyclopedia Britannica, ExploreLoneStarCoastal.com, *Fish and Wildlife News,* Georgewright.org, *Handbook of Texas Online,* JohnTedesco.net, Landscape of Ghosts, *Malabar Farm* (Bromfield and Lord), McAllenRanch.net, *New York Times, Out of the Earth* (Bromfield), *River of Dreams a History of Big Bend National Park* (Welsh), Save Buffalo Bayou, Texasbeyondhistory.net, Texas Parks and Wildlife Department, *Texas Monthly,* Texas State University Oral History Project, *The Book of Texas Bays* (Blackburn), The Conservation Fund, The History Center for Aransas County, The Nature Conservancy

Chapter 11

Interviews

Andrew Sansom

Other Sources

Blackland NPAT, *Conservation Easements: A Guide for Texas Landowners* (Francell and Ferguson), Environmental Protection Agency, Mid-Coast Chapter of Texas Master Naturalist, Smithsonian's National Zoo and Conservation Biology Institute, Texas A&M Natural Resources Institute, Texas Land Trust Council, Texas Legacy Project, *Texas Monthly*, Texas Parks and Wildlife Department, Texas Time Travel, Texasprairie.org, The Nature Conservancy, The TXGenWeb Project

Chapter 12

Interviews

Andrew Sansom, David Riskind

Other Sources

A Texas Ranching Family: The Story of E. K. Fawcett (Finnegan), *Austin Chronicle*, *Associated Press*, Briscoe Center for American History, CampFawcett.org, Clements Papers Project, *Handbook of Texas Online*, Huecotanks.com, New Mexico Land Conservancy, *New York Times*, *Nuclear Newswire*, Passport to Texas, *San Angelo Standard-Times*, Texas Escapes, *Texas Monthly*, *Texas Observer*, Texas Parks and Wildlife Department, *Texas Parks & Wildlife Magazine*, Texas State University Oral History Project, Texas Sunset Advisory Commission, *Texas Tribune*

Chapter 13

Interviews

Andrew Sansom

Other Sources

Austin Chronicle, Bizjournals.com, Common Reader, *Corpus Christi Caller-Times*, *Gulf Coast Cattleman*, *Let the People in: The Life and Times of Ann Richards* (Reid), *Lubbock Avalanche-Journal*, McIvor Ranch, National Governors Association, *Newsweek*, *New York Times*, *Northeast News*, Outdoor Industry Association, PBS, Sierra Club, *Slate*, Texas Legacy Project, Texas Parks and Wildlife Department, *Texas Parks & Wildlife Magazine*, *Texas Monthly*, *Texas Tribune*, Texas State University Oral History Project, *Texas Observer*, *Tyler Morning Telegraph*, The Free Library, *The Natural Heritage of Texas*, The Swift Lift, *Wall Street Journal*

Chapter 14

Interviews

George Bristol, Jeff Francell, Dee Halliburton, John Howard, Karl Rove, Andrew Sansom

Other Sources

American Presidency Project, *Atlantic, Dallas Morning News*, Legal Planet, *New Atlantis, Newsweek, Science, Texas Monthly, Texas Observer*, Texas Parks and Wildlife Department, The Bush School of Government and Public Service at Texas A&M University, *Vox, Washington Post, Wired*

Chapter 15

Interviews

Cindy Loeffler, Warren Pulich, Andrew Sansom

Other Sources

Aquarena Springs (Phillips), Edwardsaquifer.net, *Garden & Gun, Handbook of Texas Online*, H2O Headwaters to Ocean, Index of Texas Archaeology, *Nation*, NBC News, *San Antonio Express-News*, San Marcos River Foundation, StateImpact/NPR, *Temple Daily Telegram, Texas Co-op Power, Texas Highways*, Texas Living Waters Project, *Texas Observer, Texas Parks & Wildlife Magazine*, Texas State University, Texas State University Newsroom, Texas State University Oral History Project, *Texas Water Journal*, The Historical Marker Database, *The San Marcos: A River's Story* (Kimmel), The Meadows Center for Water and the Environment, US Geological Survey, Visit San Marcos, Waymarking.com

Chapter 16

Interviews

Bruce Esterline, Robert Mace, Carlos Rubinstein, Andrew Sansom

Other Sources

Bizjournals.com, EduRank.org, *Handbook of Texas Online*, Hill Country Alliance, H2O Headwaters to Ocean, Los Caminos del Rio Heritage Project, *MPRINT Magazine, New York Times*, Talk+Water, Texas State University, Texas State University Newsroom, Texas State University Oral History Project, The Meadows Center for Water and the Environment Annual Report 2016–2017, *University of Pennsylvania Law Review*, ULegal Inc., World Birding Center

Chapter 17

Interviews

Dan Opdyke, Andrew Sansom

Other Sources

A Thirsty Land: The Making of an American Water Crisis (McGraw), *Austin American-Statesman*, Environmental Stewardship, *Hydrological Sciences Journal*, International Network of Basin Organizations, NPR, *Of Texas Rivers and Texas Art* (Sansom and Reaves), *Oklahoma Energy Today*, sosecretoccultandconcealed.com, Texas Environmental Flows Initiative, Texas Living Waters Project, Texas Parks and Wildlife Department, *Texas Parks & Wildlife Magazine*, Texas State Library and Archives

Commission, Texas Water Development Board, Texas Water Law Institute, *Texas Water Journal*, The Nature Conservancy, University of Texas Civil Architectural and Environmental Engineering, *Water in Texas—An Introduction* (Sansom, Armitano, Wassenich), World Economic Forum

Chapter 18

Interviews

Andrew Sansom, Molly Stevens

Other Sources

Austin Monthly, Child Mind Institute, Climategrades.org, CRM.org, *Discovering Westcave: The Natural & Human History of a Hill Country Nature Preserve* (Caran and Davenport), Encyclopedia.com, Environmental Protection Agency, Greater Good, Ianbanyard.com, *Jackson Hole News & Guide*, *Miseducation: How Climate Change Is Taught in America* (Worth), National Environment Education Foundation, National Library of Medicine, National Science Foundation, *New York Times*, Onlineu.com, Penn State News, *Public School Review*, Richard Louv Blog, *Scientific American*, Texas Freedom Network, *Texas Observer*, Texas Parks and Wildlife Department, *Texas Parks & Wildlife Magazine*, Texas State University Department of Geography and Environmental Studies, *Texas Tribune*, Texas Tribune Festival, The Library of Congress, US National Library of Medicine, Visitsunsetcountry.com, *Washington Post*, Youth First

Chapter 19

Interviews

Andrew Sansom, Mollie Walton

Other Sources

Aggie Horticulture, American Academy for Park and Recreation Administration, ArcGIS Story Maps, *Austin American-Statesman*, *Austin Chronicle*, Austin History Center, Bambergerranch.org, Briscoe Center for American History Digital Collections, Canada Energy Regulator, Citizens Preserving Floyd County, *Community Impact*, Hill Country Naturalist, *Houston History Magazine*, josephjearthman.funeraltechweb.com, *Let the Water Do the Work: Induced Meandering, an Evolving Method for Restoring Incised Channels* (Zeedyk, Clothier, Gadzia), National Mississippi River Museum and Aquarium, *Nature News*, *New York Times*, Project Drawdown, Quivira Coalition, Resilience.org, *Rock & Vine*, Sage Grouse Initiative, *Seasons at Selah: The Legacy of Bamberger Ranch Preserve* (Sansom, Yates, Langford), Sierra Club, Texas A&M Agrilife Extension, Texas A&M Institute of Renewable Natural Resources, Texas Fauna Project, *Texas Highways*, Texas Hill Country Land Trust, *Texas Observer*, Texas State University Oral History Project, The Nature Conservancy, US Fish and Wildlife Service, Wildlife Habitat Federation, *Wimberley View*, World Economic Forum

Chapter 20

Interviews

Jim Blackburn, John Howard, Ken Kramer, Dan Opdyke, Alex Sansom, Andrew Sansom

Other Sources

Animalia.bio, American Association for the Advancement of Science, *American Experience* (PBS), *Associated Press*, *Audubon*, *Atlantic*, BBC News, BCarbon, Carboncredits.com, Center for American Progress, Earth.org, *Economist*, Geophysical Fluid Dynamics Laboratory, Globenewswire.com, GreatNonprofits.org, Hill Country Alliance, *Houston Chronicle*, *Lancet*, *Local Environment*, *New York Times*, Newswest9. com, PBS News Hour, Pecos County Historical Commission, Phys.org, Rice News and Media Relations, *San Antonio Business Journal*, *San Antonio Express-News*, *Society & Natural Resources*, Study.com, *Texas Monthly*, *Texas Observer*, Texas Tech University Civil Engineering, *Texas Tribune*, The Climate Reality Project, *Today in Energy*, *Tufts Now*, United Nations Sustainable Development Goals, US Department of the Interior, US Energy Information Administration, *Water in Texas: An Introduction* (Sansom, Armitano, Wassenich), World Health Organization

Selected Bibliography

Personal Interviews

Kay Bailey Hutchison—August 30, 2022

J. David Bamberger—August 6, 2021

Jim Blackburn—September 17, 2020

George Bristol—September 9, 2021

Bruce Esterline—June 6, 2022

Jeff Francell—March 1, 2022

Dee Halliburton—March 15, 2022

John Howard—May 3, 2022

Ken Kramer—May 4, 2021

Cindy Loeffler—August 4, 2022

Robert Mace—June 6, 2022

Dan Opdyke—May 12, 2022

Warren Pulich—June 28, 2022

David Riskind—November 1, 2022

Karl Rove—March 15, 2022

Carlos Rubinstein—September 22, 2022

Alex Sansom—December 8, 2022

April Sansom—June 27, 2022

Andrew Sansom—April 7, 2020; April 21, 2020; May 5, 2020; May 19, 2020; June 2, 2020; June 30, 2020; July 14, 2020; September 1, 2020; August 18, 2020; December 3, 2020; October 13, 2022; December 17, 2020; November 23, 2021; January 27, 2022; February 21, 2022; March 15, 2022; April 23, 2022; June 7, 2022; June 10, 2022; July 6, 2022; July 8, 2022; August 14, 2022; August 18, 2022; September 1, 2022; September 14, 2022; November 7, 2022; November 7, 2022

Andrew Sansom Jr.—November 11, 2022

Nona Sansom—October 6, 2020

Molly Stevens—November 9, 2022

Mollie Walton—August 10, 2022

Books

Aulbach, Louis F. *Buffalo Bayou: An Echo of Houston's Wilderness Beginnings.* Houston: L. F. Aulbach, 2012. http://www.epperts.com/lfa/BB63.html.

Blackburn, Jim. *The Book of Texas Bays.* College Station: Texas A&M University Press, 2014.

Bromfield, Louis. *Out of the Earth*. New York: Harper and Brothers, 1950.

Bromfield, Louis, and Kate Lord. *Malabar Farm*. New York: Ballantine Books, 1970.

Caran, S. Christopher, and Elaine Davenport. *Discovering Westcave: The Natural and Human History of a Hill Country Nature Preserve*. College Station: Texas A&M University Press, 2016.

Finnegan, John. *A Texas Ranching Family: The Story of E. K. Fawcett*. Bloomington, IN: AuthorHouse, 2007.

Fritz, Edward C., and Jess Alford. *Realms of Beauty: The Wilderness Areas of East Texas*. Austin: University of Texas Press, 1986.

Kimmel, Jim. *The San Marcos: A River's Story*. College Station: Texas A&M University Press, 2006.

McGraw, Seamus. *A Thirsty Land: The Making of an American Water Crisis*. Austin: University of Texas Press, 2018.

Reid, Jan. *Let the People in. The Life and Times of Ann Richards*. Austin: University of Texas Press, 2014.

Sansom, Andrew, Emily R. Armitano, and Tom Wassenich. *Water in Texas: An Introduction*. Austin: University of Texas Press, 2008.

Sansom, Andrew, and Clemente Guzman. *Scout, the Christmas Dog*. College Station: Texas A&M University Press, 2006.

Sansom, Andrew, and William E. Reaves. *Of Texas Rivers & Texas Art*. College Station: Texas A&M University Press, 2017.

Sansom, Andrew, Jan Reid, and Wyman Meinzer, *Texas Lost: Vanishing Heritage*. Dallas: Parks and Wildlife Foundation of Texas, 1995.

Sansom, Andrew, Rusty Yates, and David K. Langford. *Seasons at Selah: The Legacy of Bamberger Ranch Preserve*. College Station: Texas A&M University Press, 2018.

Welsh, Michael E., *Landscape of Ghosts, River of Dreams: A History of Big Bend National Park*. Washington, DC: National Park Service, U.S. Department of the Interior, 2002. http://npshistory.com/publications/bibe/adhi-2008.pdf

Steffy, Loren C. *George P. Mitchell: Fracking, Sustainability, and an Unorthodox Quest to Save the Planet*. College Station: Texas A&M University Press, 2019.

Worth, Katie. *Miseducation: How Climate Change Is Taught in America*. New York: Columbia Global Reports, 2021.

Zeedyk, William D., Van Clothier, and Tamara E. Gadzia, *Let the Water Do the Work: Induced Meandering, an Evolving Method for Restoring Incised Channels*. White River Junction, VT: Chelsea Green Publishing, 2009.

News Outlets

"Accident Cools Down Campaign In Austin on Atomic Plant Bonds," *New York Times*, April 5, 1979, https://www.nytimes.com/1979/04/05/archives/accident-cools-down-campaign-in-austin-on-atomic-plant-bonds.html.

"Anne Morton Kimberly Dies at Her Easton Home," *Star Democrat*, September 11, 2011, https://www.stardem.com/obituaries/anne-morton-kimberly-dies-at-her-easton-home/article_23b9ba32–20e2–5fd8–93dd-b750abacf64a.html.

"Austin City Council Election 1979," *Austin Bulldog*, n.d., https://theaustinbulldog. org/austin-city-council-election-1979/.

Barer, David, "What Are 'Environmental Flows' and How Does Texas Protect Them?" *StateImpact NPR*, January 16, 2013, https://stateimpact.npr.org/texas/2013/01/16/ what-are-environmental-flows-and-how-does-texas-protect-them/.

Barta, Carolyn. "How George H. W. Bush Turned Texas Red," *Dallas Morning News*, December 5, 2018, https://www.dallasnews.com/opinion/commentary/2018/12/ 05/how-george-h-w-bush-turned-texas-red.

Bernd, Candice. "The Long Battle to Stop the Kinder Morgan Pipeline," *Texas Observer*, November 5, 2019, https//www.texasobserver.org/kinder-morgan-texas-hill-country-lawsuit.

"Best Architect: Emily Little" *Austin Chronicle*, 2007, https://www.austinchronicle. com/ best-of-austin/year: 2007/poll: readers/category: architecture-and-lodging/ emily-little-best-architect/.

"Biden Admin Halts Oil Drilling in Alaska Wildlife Refuge," *Associated Press* via *Voice of America News*, June 2, 2021, https://www.voanews.com/a/economy-business_ biden-admin-halts-oil-drilling-alaska-wildlife-refuge/6206510.html.

Blakeslee, Nate. "Naked City Sun Sets Slowly," *Austin Chronicle*, June 9, 2000, https:// www.austinchronicle.com/news/2000–06–09/77541/.

Blackburn, Jim, and Susan Combs. "Opinion: How Texas farmers can profit while pumping carbon into the soil." *Houston Chronicle*, June 6, 2022, https://www. houstonchronicle.com/opinion/outlook/article/Opinion-How-Texas-farmers-can-profit-while-17219523.php.

Blanks, Annie. "The Cautionary Tale of the San Marcos Gambusia, Declared Extinct in Fast-Growing Hays County," *San Antonio Express-News*, October 11, 2021, https://www.expressnews.com/news/local/article/San-Marcos-gambusia-extinct-16519656.php.

Bohnen, Jerry. "Texas to Begin Campaign to Conserve Water," *Oklahoma Energy Today*, July 16, 2018, http://www.okenergytoday.com/2018/07/texas-to-begin-campaign-to-conserve-water.

Breuninger, Kevin. "Former President George W. Bush Will Fundraise for Liz Cheney as Trump and GOP Rivals Target Her House Seat," *CNBC*, September 23, 2021, https://www.cnbc.com/2021/09/22/george-w-bush-to-fundraise-for-liz-cheney-as-trump-and-gop-rivals-target-her-seat-.html.

Brodesky, Josh. "Why can't Greg Abbott accept man-made climate change?" *San Antonio Express-News*, March 9, 2019, https://www.mysanantonio.com/opinion/ columnists/josh_brodesky/article/Why-can-t-Greg-Abbott-accept-man-made-climate-13674643.php.

Brody,   Jane E. "Snow Geese Survive All Too Well, Alarming Conservationists," *New York Times*, February 11, 1997, https://www.nytimes.com/1997/02/11/ science/snow-geese-survive-all-too-well-alarming-conservationists.html.

Brown, Patricia Leigh. "Mr. Fisk Builds His Green House," *New York Times*, February 15, 1996, https://www.nytimes.com/1996/02/15/garden/mr-fisk-builds-his-green-house.html.

"Caprock Chronicles: Saving the Bison on the Texas High Plains," *Lubbock Avalanche-Journal*, January 23, 2021, https://www.lubbockonline.com/story/news/history/2021/01/22/caprock-chronicles-saving-bison/6682031002/.

Bryce, Robert. "Wild, Wild Wildlife: Texas Parks & Wildlife's Andy Sansom," *Austin Chronicle*, May 9, 1997, https://www.austinchronicle.com/news/1997-05-09/528161.

Cockerham, Sean. "Alaska's Walter Hickel, Who Nixon Fired over Vietnam, Dies," *Anchorage Daily News*, April 8, 2010, https://www.mcclatchydc.com/news/politics-government/article24582163.html.

Collier, Kiah. "Report: Texas Ranks Second in Budget Cuts for Environmental Protection," *Texas Tribune*, December 5, 2019, https://www.texastribune.org/2019/12/05/report-texas-ranks-second-budget-cuts-environmental-protection.

Coppedge, Clay. "Aquarena Center Both Educational and Fun," *Temple Daily Telegram*, May 19, 2003, https://www.tdtnews.com/archive/article_6ab31301-232b-5132-ac81-0f1c75422358.html.

Corso, Jessica. "SAWS Turns on Tap of $900M Vista Ridge Project," *San Antonio Business Journal*, May 6, 2020, https://www.bizjournals.com/sanantonio/c/saws-turns-on-tap-of-900m-vista-ridge-project.html.

Daly, Matthew. "Biden Promises to Take Action on Climate Change despite Setbacks," *PBS News Hour*, July 16, 2022, https://www.pbs.org/newshour/nation/biden-plans-to-use-executive-action-to-combat-climate-action-despite-setbacks.

Davenport, Coral. "Restoring Environmental Rules Rolled Back by Trump Could Take Years." *New York Times*, January 22, 2021, https://www.nytimes.com/2021/01/22/climate/biden-environment.html.

Dingfelder, Sadie. "The Isolation of the Pandemic Caused Her to Form a New and Intense Relationship to Nature. She Was Hardly Alone." *Washington Post*. December 28, 2020, https://www.washingtonpost.com/magazine/2020/12/28/isolation-pandemic-caused-her-form-new-intense-relationship-nature-she-was-hardly-alone.

Douglas, Erin. "Climate Change Is Making Texas Hotter, Threatening Public Health, Water Supply and the State's Infrastructure," *Texas Tribune*, October 7, 2021, https://www.texastribune.org/2021/10/07/texas-climate-change-heat-water.

Dougherty, Philip H. "Advertising," *New York Times*, June 18, 1975, https://www.nytimes.com/1975/06/18/archives/advertising-is-fea-or-its-agency-fuelish.html.

Douglas, Erin. "San Antonio Built a Pipeline to Rural Central Texas to Increase Its Water Supply. Now Local Landowners Say Their Wells Are Running Dry" *Texas Tribune*, August 2, 2021, https://www.texastribune.org/2021/08/02/san-antonio-water-supply-rural-wells.

Duff, Audrey. "Cowboys & Critters," *Austin Chronicle*, January 1996, https://www.austinchronicle.com/news/1996–01–26/530503/.

Eisenbaum, Joel. "Pollution Enforcement: TCEQ Sides with Power Plants." *KPRC TV*, July 1, 2022, https://www.click2houston.com/news/investigates/2022/07/01/pollution-enforcement-tceq-sides-with-power-plants.

"EPCOR USA Selected as Operator of Vista Ridge Project." *GlobeNewswire*, November 16, 2018, https://www.globenewswire.com/en/news-release/2018/11/16/1653055/0/en/EPCOR-USA-Selected-as-Operator-of-Vista-Ridge-Project.html.

Fehling, Dave. "Texas' Lax Pollution Enforcement Leads Harris County to Take Action," *StateImpact NPR*, November 8, 2011, https://stateimpact.npr.org/texas/2011/11/08/harris-county-attorneys-office-on-tceq-offensive-sue-polluters/.

"Former Congressman Pete McCloskey Authors New Book Titled The Story of the First Earth Day—How Grassroots Activism Changed the World," *Business Wire*, April 21, 2020, https://www.businesswire.com/news/home/20200421005879/en/Former-Congressman-Pete-McCloskey-Authors-New-Book-Titled-The-Story-of-the-First-Earth-Day—How-Grassroots-Activism-Changed-the-Worldan.

Galbraith, Kate. "Ken Kramer: The TT Interview," *Texas Tribune*, April 3, 2012, https://www.texastribune.org/2012/04/03/texas-sierra-clubs-outgoing-head-reflects-changes.

Garcia, Gilbert. "Garcia: Former Councilman Will Run as an Independent for Vacant Commissioners Court Seat," *San Antonio Express News*, December 21, 2021, https://www.expressnews.com/columnist/gilbert-garcia/article/Garcia-Former-councilman-will-run-as-an-16720905.php.

Hamilton, Reeve, and Jay Root. "A Governor's Long Legacy of Brashness and Brawls," *New York Times*, December 27, 2014, https://www.nytimes.com/2014/12/28/us/a-governors-long-legacy-of-brashness-and-brawls.html.

Henry, Terrence. "Too Many Straws in the Ground: An Interview with Andrew Sansom," *StateImpact NPR*, March 28, 2012, https://stateimpact.npr.org/texas/2012/03/28/too-many-straws-in-the-ground-an-interview-with-andrew-sansom.

"Heritage Lost," *Austin Chronicle*, 1995, https://www.austinchronicle.com/news/1995-10-13/529912.

Hershey, Terry. "In Praise of a Praiser." *New York Times*, May 13, 1993, https://www.nytimes.com/1993/05/13/garden/l-in-praise-of-a-praiser-930893.html.

Hevesi, Dave. "Edward H. Harte, Texas Newspaper Executive, Dies at 88." *New York Times*, May 24, 2011, https://www.nytimes.com/2011/05/24/business/24harte.html.

Hooks, Christopher. "Burning Down The House: Joe Straus and the End of the Moderate Texas Republican," *Texas Observer*, October 25, 2017, https://www.texasobserver.org/burning-house-joe-straus-end-moderate-texas-republican.

Jimenez, Jesus. "Texas Scientists Offer to Brief Gov. Greg Abbott on Climate Change, Railroad Commissioner Says It's 'Far from Settled Science,'" *Dallas Morning News*, January 9, 2019, https://www.dallasnews.com/news/environment/2019/01/09/texas-scientists-offer-to-brief-gov-greg-abbott-on-climate-change-railroad-commissioner-says-it-s-far-from-settled-science/.

Jones, Kathryn. "Slipping in and out of Mexico," The New York Times, January 26, 1997, https://www.nytimes.com/1997/01/26/travel/slipping-in-and-out-of-mexico.html.

Jose, Katharine. "Judge Dismisses Claims against Texas Railroad Commission, Kinder Morgan in Permian Highway Pipeline Lawsuit," *Community Impact*, June 26, 2019, https://communityimpact.com/austin/san-marcos-buda-kyle/editors-pick/2019/06/26/judge-dismisses-claims-against-texas-railroad-commission-kinder-morgan-in-permian-highway-pipeline-lawsuit.

Kaspar, Mary Alice, "School to Study Water," *Austin Business Journal*, May 12, 2002, https://www.bizjournals.com/austin/stories/2002/05/13/story4.html.

Kilday, Anne Marie. "Faces in the Crowd: Jim Blackburn," *Houston Chronicle*, July 21, 2011, https://www.chron.com/neighborhood/heights-news/article/Faces-in-the-Crowd-Jim-Blackburn-1486251.php.

Knight, Steve. "Pay to Play: TPWD Institutes Fee for Managed Lands Deer Programs," *Tyler Morning Telegraph*, May 2021, https://tylerpaper.com/opinion/columnists/pay-to-play-tpwd-institutes-fee-for-managed-lands-deer-programs/article_474359de-b2ca-11eb-b585-ab318e6896d0.html.

Lackey, Jerry. "Homestead: Herders Leave Mark on Dolan Creek." *San Angelo Standard-Times*, March 3, 2012, https://archive.gosanangelo.com/business/homestead-herders-leave-mark-on-dolan-creek-ep-439374108–356373821.html.

Lee, Victoria, "Parks See Busiest Winter in Decades," *Jackson Hole News & Guide*, February 3, 2021, https://www.jhnewsandguide.com/news/business/parks-see-busiest-winter-in-decades/article_db5e71b9-ff46-59b8-9d33-7facdad8b91d.html.

LeBlanc, Pam. "Nature Conservancy Protects Fragile Coastal Lands at Mad Island Marsh," *Austin American-Statesman*, May 5, 2017.

LeBlanc, Pamela. "Ranchland in Recovery," Austin American-Statesman, June 7, 2009, https://www.txst.edu/mathworks/about/news/bamberger09.html.

Lopez, Brian. "Texas Education Board Considers How Middle Schools Teach Climate Change and Sexuality as Officials Fight over Library Books," *Texas Tribune*, November 16, 2021, https://www.texastribune.org/2021/11/16/texas-education-board-climate-change-sexuality.

Martin, Douglas. "Robert O. Anderson, Oil Executive, Dies at 90," *New York Times*, Dec. 6, 2007. https://www.nytimes.com/2007/12/06/business/06anderson.html.

Maxa, Rudy. "Charles Wilson," *Washington Post*, November 5, 1978, https://www.washingtonpost.com/archive/lifestyle/magazine/1978/11/05/charles-wilson/d045ea63-b40e-46a3-9343-70e60b43ba40.

Mazur, Jeremy. "Increasing Drought Conditions Threaten Texas' Future Prosperity," *Katy Times*, June 23, 2022, http://katytimes.com/stories/increasing-drought-conditions-threaten-texas-future-prosperity,7871.

McCrimmon, Ryan. "Former Land Commissioner Bob Armstrong Dies at 82," *Texas Tribune*, March 2, 2015, https://www.texastribune.org/2015/03/02/former-land-commissioner-bob-armstrong-dies-82.

McGrath, Matt. "Climate Change: IPCC Report Warns of 'Irreversible' Impacts of Global Warming," *BBC News*, February 28, 2022, https://www.bbc.com/news/science-environment-60525591.

McLeod, Gerald E., "Day Trips," *Austin Chronicle*, October 20, 2000, https://www
.austinchronicle.com/columns/2000-10-20/78991.

McWilliams, Colton. "Battle for the Heart of Texas," *Wimberley View*, May 13,
2022, https://www.wimberleyview.com/features/%E2 80 98battle-heart-texas
E280%99.

Mendez, Maria, and Erin Douglas. "Seven Ways Climate Change Is Already Hitting
Texans," *Texas Tribune*, May 18, 2022, https://www.texastribune.org/2022/05/
18/climate-change-texas/.

Merks, Kelly. "Parks Department Set to Decide on Devil's River Purchase," *Texas
Tribune*, December 7, 2010, https://www.texastribune.org/2010/12/06/parks-
department-will-vote-on-devils-river-land.

Mooney, Chris, and Harry Stevens, "The US Plan to Avoid Extreme Climate
Change Is Running out of Time," *Washington Post*, July 19, 2022, https://www
.washingtonpost.com/climate-environment/2022/07/18/climate-change-
manchin-math.

Moritz, John C. "Moritz: How Texas Went from Having Weak Governors to
Powerful Governors over Past 24 Years," *Corpus Christi Caller-Times*, October
23, 2021, https://www.caller.com/story/news/columnists/john-moritz/2021/
10/22/greg-abbott-continues-recent-trend-expanding-power-governors/
6133480001.

Murphy, Ryan. "Interactive Map: Texas Cities at Risk of Running Out of Water," *Texas
Tribune*, November 13, 2011, https://flatpage-archive.texastribune.org/library/
data/tceq-high-priority-water-locations.

Newport, Frank. "Is George W. Bush Riding His Father's Coattails?," *Gallup*, July
1, 1998, https://news.gallup.com/poll/4600/George-Bush-Riding-Fathers-
Coattails.aspx.

"No Republican Senator Supported a Climate Plan—Where Is the Party on the
Issue?" *Guardian*, July 22, 2022, https://www.theguardian.com/us-news/2022/
jul/22/republican-party-climate-issue.

O'Connor, Anahad. "The Claim: Exposure to Plants and Parks Can Boost Immunity,"
New York Times, July 5, 2010, https://www.nytimes.com/2010/07/06/health/
06real.html.

Oko, Dan. "A Shifting Landscape," *Austin Chronicle*, February 2002, https://www.
austinchronicle.com/news/2002-02-08/84598.

Pearce, Michael Pearce. "Too Many Geese for Their Own Good," *Wall Street Journal*,
April 8, 1999), https://www.wsj.com/articles/SB923529277150517864.

"Perry's Commissioners," *Austin Chronicle*, February 8, 2002, https://www
.austinchronicle.com/news/2002-02-08/84601/.

Piasecki, Mary Alice. "Francell Takes New Job," *Austin Business Journal*, January 2001,
https://www.bizjournals.com/austin/stories/2001/01/15/story6.html.

Polasek, Jonathan. "Midland Reaches Water Deal Covering the next 50 Years,"
KWES-TV Midland, May 13, 2020, https://www.newswest9.com/article/news/

local/midland-water-deal/513-60279ab1-73db-49cc-9637-4fd53e972348.

Pressley, Sue Anne. "Bush Defeats Richards for Texas Governorship," *Washington Post*, November 9, 1994, https://www.washingtonpost.com/archive/politics/1994/11/09/bush-defeats-richards-for-texas-governorship/e44870f9-7797-450a-9288-f7d0ce53fb24/.

Price, Asher. "Hoping to Raise Water Awareness, State Returns to a Famous Ad Man," Austin American-Statesman, September 25, 2018, https://www.statesman.com/story/news/2018/07/14/hoping-to-raise-water-awareness-state-returns-to-a-famous-ad-man/10152483007.

"Proposition 8 Will Give a Big Boost to State Parks, Youth and Outdoors," *Northeast News*, October 23, 2001, https://www.nenewsroom.com/2001/10/proposition-8-will-give-a-big-boost-to-state-parks-youth-and-outdoors.

Ramsey, Ross. "Bigger than Life," *Texas Tribune*, September 18, 2006, https://www.texastribune.org/2006/09/18/bigger-than-life.

Samenow, Jason. "Unforgiving Heat Wave in Texas and Southern Plains to Worsen Next Week," *Washington* Post, July 15, 2022. https://www.washingtonpost.com/climate-environment/2022/07/14/texas-oklahoma-heat-wave-plains.

Salmi, Noelle Alejandra. "There Are Too Many Environmental Organizations," *Matador Network*, August 16, 2022, https://matadornetwork.com/read/too-many-environmental-organizations.

Sandomir, Richard. "Nathaniel Reed, 84, Champion of Florida's Environment, Is Dead," *New York Times*, July 13, 2018, https://www.nytimes.com/2018/07/13/obituaries/nathaniel-reed-84-champion-of-floridas-environment-is-dead.html.

Sansom, Andrew, "How to Save Texas' Crumbling Parks," *TribTalk*, May 2015, https://www.tribtalk.org/2015/05/14/how-to-save-texas-crumbling-parks.

Saxon, Wolfgang. "Algur Hurtle Meadows, at 79; Turned Oil Concern into Empire," *New York Times*, June 13, 1978, https://www.nytimes.com/1978/06/13/archives/algur-hurtle-meadows-at-79-turned-oil-concern-into-empire-two.html.

Schechter, David. "Texans Face Greater Risk of Heat, Drought and Hurricanes, but Abbott Administration Has No Plan to Tackle Future Threats of Climate Change." *ABC News 8 WFAA*, November 7, 2021, https://www.wfaa.com/article/tech/science/climate-change/texas-climate-change-risks-gov-abbott-un-summit-scotland/287-75b20337-0f82-4f83-a214-09b2f07309be.

Smith, Amy. "Naked City," *Austin Chronicle*, February 2001, https://www.austinchronicle.com/news/2001-02-02/80442.

Sottile, Chiara A. "Flying Pigs and Mermaids: Crumbled Theme Park Reborn as Nature Center," *NBC News*, https://www.nbcnews.com/business/business-news/flying-pigs-mermaids-crumbled-theme-park-reborn-nature-center-n555931.

Suro, Roberto. "The 1990 Elections: Texas; Richards Promises a New Direction," *New York Times*, November 8, 1990), https://www.nytimes.com/1990/11/08/us/the-1990-elections-texas-richards-promises-a-new-direction.html.

"Texas Construction Firm to Pay $750 Million: Suit Over Nuclear Project Settled," *Los Angeles Times*, May 31, 1955, https://www.latimes.com/archives/la-xpm-1985-05-31-fi-14780-story.html.

"Texas—How Texas Became a 'Red' State," *PBS Frontline*, April 2005, https://www
.pbs.org/wgbh/pages/frontline/shows/architect/texas/realignment.html.

"Texas Is Facing Its Worst Drought since 2011. Here's What You Need to Know,"
Prosper Press, August 24, 2022, https://www.prosperpressnews.com/2022/08/
24/texas-is-facing-its-worst-drought-since.

"Theodore Roosevelt and the Environment," *PBS / American Experience*, https://
www.pbs.org/wgbh/americanexperience/features/tr-environment.

Tomlinson, Chris. "Climate Change Is Taking Thousands of Lives—and Could
Suck $178 Trillion from the Global Economy," *Houston Chronicle*, July 25, 2022,
https://www.houstonchronicle.com/business/columnists/tomlinson/article/
climate-change-is-causing-problems-across-globe-17322322.php.

Tomlinson, Chris. "Fighting Climate Change Requires Changing Texas Beef and Oil
Culture," *Houston Chronicle*, October 25, 2021, https://www.houstonchronicle.
com/business/columnists/tomlinson/article/beef-oil-culture-texas-climate-
change-fight-16545487.php.

Totenberg, Nina. "Supreme Court Restricts the EPA's Authority to Mandate Carbon
Emissions Reductions," *NPR*, June 30, 2022, https://www.npr.org/2022/06/30/
1103595898/supreme-court-epa-climate-change.

Watkins, Katie, "Rice Launches Climate Change Initiative with $10 Million
Donation from Shell," *Houston Public Media*, December 10, 2019, https://www
.houstonpublicmedia.org/articles/news/energy-environment/2019/12/10/
353662/rice-launches-climate-change-initiative-with-10-million-donation-
from-shell.

"William B. Pond," Legacy.com and *The Sacramento Bee*, September 23, 2009,
https://www.legacy.com/us/obituaries/sacbee/name/william-pond-obituary?
id=12475475.

Weise, Elizabeth. "The World Is 'Perilously Close' to Irreversible Climate Change.
5 Tipping Points Keep Scientists up at Night," *Phys.org*, April 7, 2022, https://
phys.org/news/2022-04-world-perilously-irreversible-climate-scientists.html.

Podcasts

Saleh, Nivien. "The Next Houston Hurricane May Be Worse Than Harvey—Let's
Get Ready!" Interview with Terence O'Rourke." *Houston & Nature*, July 1,
2020. https://houstonnature.com/wp-content/uploads/2020/07/HN05-Houston
-hurricane-Terence-ORourke-Transcript.pdf.

Votteler, Todd. "#48—Andrew Sansom—Using Water Markets to Benefit the Texas
Environment." Interview with Andrew Sansom. *Talk + Water*, October 25,
2022, https://podcasts.apple.com/us/podcast/48-andrew-sansom-using-water
-markets-to-benefit-the/id1450729314?i=1000583882301.

Magazines and Journals

Adler, Jonathan H. "The Conservative Record on Environmental Policy," *The New
Atlantis*, No. 39, Summer 2013. https://www.thenewatlantis.com/publications/
the-conservative-record-on-environmental-policy.

Ahmed, Amal. "Texas Students Are Receiving a Miseducation on Climate Change." *The Texas Observer*, November 17, 2021. https://www.texasobserver.org/texas-students-are-receiving-a-miseducation-on-climate-change/.

Allmon, Eric, and David Frederick. "A Defense of the Contested Case Hearing Process for Texas Commission on Environmental Quality Environmental Permit Decisions." *Texas Environmental Law Journal*, 44 (n.d.): pp. 175–234. http://www.texascenter.org/publications/TELJ%20Allmon%20CCH%20Article.pdf.

"American Climate Policy Is in Tatters." *Economist*, July 21, 2022, https://www.economist.com/united-states/2022/07/21/american-climate-policy-is-in-tatters.

Banks, Suzy. "Winging It." *Texas Monthly*, September 1997. https://www.texasmonthly.com/travel/winging-it/.

Barron, Colette. "In Water We Trust." *Texas Parks & Wildlife Magazine*, July 2006. https://tpwmagazine.com/archive/2006/jul/scout1/.

Beckner, Sydney, Wendy Jepson, Christian Brannstrom, and John Tracy. "'The San Antonio River Doesn't Start in San Antonio, It Now Starts in Burleson County': Stakeholder Perspectives on a Groundwater Transfer Project in Central Texas." *Society & Natural Resources* 32, no. 11 (August 5, 2019): pp. 1222–1238. https://doi.org/10.1080/08941920.2019.1648709.

Beyle, Thad. "Is George W. Bush a 'Weak' Governor?" *Slate*, January 5, 2000. https://slate.com/news-and-politics/2000/01/is-george-w-bush-a-weak-governor.html.

Brandimarte, Cynthia. "Park Pick: Big Bend's Secret Waterfall. Madrid Falls Is Big Bend Ranch's Loveliest Surprise." *Texas Parks & Wildlife Magazine*, 2017. https://tpwmagazine.com/archive/2017/aug/scout4_parkpick_bigbend/.

Burka, Paul. "Ann of A Hundred Days." *Texas Monthly*, May 1991. https://www.texasmonthly.com/news-politics/ann-richards-first-hundred-days/.

"Bush Pushes Environment Platform." *Wired*, June 2, 2000. https://www.wired.com/2000/06/bush-pushes-environment-platform/.

Cartwright, Gary. "Shooting Blanks," *Texas Monthly*, December 1, 1996. https://www.texasmonthly.com/news-politics/shooting-blanks-2/.

Curtis, Gregory. "Fix the Roof." *Texas Monthly*, March 1999. https://www.texasmonthly.com/news-politics/fix-the-roof/.

Clark, Emily Suzanne. "Rough and Religious." *Common Reader*, January 16, 2020. https://commonreader.wustl.edu/c/rough-and-religious/.

Daniel R. Opdyke, Edmund L. Oborny, Samuel K. Vaugh & Kevin B. Mayes. "Texas environmental flow standards and the hydrology-based environmental flow regime methodology." *Hydrological Sciences Journal* 59, no. 3–4 (2014): 820–830. DOI: 10.1080/02626667.2014.892600.

Delkic, Melina. "George W. Bush 'Chased a Lot of Pussy' in His Youth." *Newsweek*, November 10, 2017. https://www.newsweek.com/george-w-bush-chased-pussy-drank-whiskey-708653.

Elkind, Peter. "The Bucks Stop Here." *Texas Monthly*, January 1990. https://www.texasmonthly.com/travel/texas-parks-and-wildlife-department-scandals/.

Farber, Dan. "Barry Goldwater, Environmentalist." *Legal Planet*, June 27, 2016. https://legal-planet.org/2016/06/27/barry-goldwater-environmentalist/.

Gaskill, Melissa. "Thirst for Knowledge." *Texas Co-op Power*, January 2019. https://texascooppower.com/thirst-for-knowledge/.

Granville Sheehy, Sandy. "Welcome to the Family." *Texas Monthly*, Feb. 1988.

Gwynne, S.C. "Genius." *Texas Monthly*, January 21, 2013. https://www.texasmonthly.com/news-politics/genius/.

Hart, Patricia Kilday. "25 Stories about Bob Bullock." *Texas Monthly*, January 20, 2013. https://www.texasmonthly.com/news-politics/25-stories-about-bob-bullock/.

Harvey, Tom. "Devils Advocates." *Texas Parks & Wildlife Magazine*, August/September 2018. https://tpwmagazine.com/archive/2018/aug/ed_3_devilsriver/index.phtml#top.

Hawke, Ethan. "Where I'm From." Edited by Pamela Colloff. *Texas Monthly*, October 2005. https://www.texasmonthly.com/articles/ethan-hawke/.

"Health in a World of Extreme Heat." *Lancet*, August 21, 2021. https://www.thelancet.com/journals/lancet/article/PIIS0140-6736(21)01860-2/fulltext.

Hodge, Larry D. "Why State Parks Offer Hunts." *Texas Parks & Wildlife Magazine*, November 2003, https://tpwmagazine.com/archive/2003/nov/scout2/.

Holley, Peter. "Why One Expert Predicts a Major Hurricane Hitting Houston Would Be 'America's Chernobyl.'" *Texas Monthly*, August 21, 2020. https://www.texasmonthly.com/news-politics/houston-hurricane-ship-channel-orourke/.

Holtcamp, Wendee. "Larger Than Life: The Inimitable Edward 'Ned' Fritz Changed the Face of Texas Conservation." *Texas Parks & Wildlife Magazine*, August 2009. https://tpwmagazine.com/archive/2009/aug/legend/.

Jarboe Russell, Jan. "Meet the Governor: Clayton Williams." *Texas Monthly*, October 1990. https://www.texasmonthly.com/news-politics/meet-the-governor-clayton-williams/.

Jefferson, John. "The Golden Age of Park Acquisitions: A Perfect Storm of Money and Support Resulted in the Purchase of Beloved State Parks." *Texas Parks & Wildlife*, August 2011. https://tpwmagazine.com/archive/2011/aug/ed_1_stateparks/.

Kaiser, Ronald. "Who Owns the Water?" *Texas Parks & Wildlife Magazine*, July 2005. http://www.texasthestateofwater.org/screening/html/waterlaws.htm.

Kreitner, Richard. "December 3, 1964: Mass Arrests of Students at the University of California, Berkeley." *Nation*, December 3, 2015. https://www.thenation.com/article/archive/december-3-1964-mass-arrests-of-students-at-university-of-california-berkeley/.

LeBlanc, Pam. "The Magical Days of Aquarena Springs in San Marcos." *Texas Highways*, July 29, 2021. https://texashighways.com/culture/history/the-magical-days-of-aquarena-springs-in-san-marcos/.

Lotz, CJ. "A Dip into Southern Mermaid History." *Garden & Gun*, July 30, 2018. https://gardenandgun.com/articles/dip-southern-mermaid-history/.

"Matagorda Island to Join Refuge System." *Fish and Wildlife News*, 1987. play.google.com/books/reader? id=sadvodokbwgC&pg=GBS.RA5-PA8&printsec=frontcover.

McNeely, Dave. "The Man Who Ran Texas." *Texas Observer*, April 19, 2019. https://www.texasobserver.org/2657-the-man-who-ran-texas-excerpts-from-a-new

-biography-of-bob-bullock-one-of-the-most-powerful-feared-and-unpredictable
-politicians-in-texas-history/.

Meyer, Robinson. "A Republican Congress Is Coming for Biden's Climate Wins." *The Atlantic*, November 2, 2022. https://www.theatlantic.com/science/archive/2022/11/climate-change-republican-congress-2022-midterms/671972/.

Mongillo, Anna. "A Love Letter to Magnolia Cafe West." *Austin Monthly Magazine*, May 6, 2020. https://www.austinmonthly.com/a-love-letter-to-magnolia-cafe-west/.

Nichols, John. "The Once Common Republican Environmentalist Is Virtually Extinct." *Nation*, August 13, 2019. https://www.thenation.com/article/archive/republican-environmentalist-virtually-extinct/.

Nietzel, Michael T. "American Colleges Facing Threats from Republicans' Turn Against Science." *Forbes*, July 18, 2021. https://www.forbes.com/sites/michaeltnietzel/2021/07/18/american-higher-ed-facing-brunt-of-gop-turn-against-science/?sh=38c0e0df3aae.

Oppenheimer, Daniel. "Bamberger: A Strongly Rooted Legacy." *Rock & Vine Magazine*, April 9, 2020. https://rockandvinemag.com/2020/04/bamberger/.

Patoski, Joe Nick. "Is Texas' Overcrowded, Underfunded State Parks System Being Loved to Death?" *The Texas Observer*, December 17, 2018. https://www.texasobserver.org/is-texas-overcrowded-underfunded-state-parks-system-being-loved-to-death/.

Patoski, Joe Nick. "Playing By the Rule: Groundwater Is Covered by an Archaic Law That Could Leave Us High and Dry." *Texas Observer*. Accessed June 24, 2010. https://www.texasobserver.org/playing-by-the-rule/.

Poole, Claire. "Whatever Happened to Walter Mischler." *Texas Monthly*, 1999. https://www.texasmonthly.com/articles/whatever-happened-to-walter-mischer/.

Ridge, Tom. "My Fellow Conservatives Are out of Touch on the Environment." *Atlantic*, April 22, 2020. https://www.theatlantic.com/ideas/archive/2020/04/environment-gop-out-touch/610333/.

Reinert, Al. "The Big Thicket Tangle." *Texas Monthly*, July 1973. https://www.texasmonthly.com/news-politics/the-big-thicket-tangle/.

Robinson, Meyer. "A Republican Congress Is Coming for Biden's Climate Wins." *The Atlantic*, November 2, 2022. https://www.theatlantic.com/science/archive/2022/11/climate-change-republican-congress-2022-midterms/671972/.

Roe, Russell. "My Breakfast with Andy." *Texas Parks & Wildlife Magazine*, October 2020. https://tpwmagazine.com/archive/2020/oct/ed_1_magnolia/index.phtml.

Rubinstein, Carlos, Curtis Seaton, and Robert E. Mace. "Beyond Senate Bill 3: How to Achieve Environmental Flows in Texas Under Prior Appropriation." *Texas Water Journal* 13, no. 1 (2022): 13–26. https://doi.org/10.21423/twj.v13i1.7115.

Sansom, Andrew. "Keeping Rivers Flowing." *Texas Parks & Wildlife Magazine*, July 2011. https://tpwmagazine.com/archive/2011/jul/ed_4_rivers/index.phtml.

Sansom, Andrew. "One Man's Half-Century Project to Heal a Hill Country Landscape Created a Legacy Reaching Far beyond His Fenceline." *Texas Highways*,

May 2019. https://texashighways.com/culture/people/one-mans-half-century-project-to-heal-a-hill-country-landscape-created-a-legacy-reaching-far-beyond-his-fenceline/.

Sansom, Andrew. "Toil and Trouble at the South Texas Nuke." *Texas Observer*, July 13, 1979. https://archives.texasobserver.org/issue/1979/07/13#page=1.

Sansom, Lindsay C. "The Transformation of Aquarena." *Texas Parks & Wildlife Magazine*, May 2013. https://tpwmagazine.com/archive/2013/may/ed_3_aquarena/index.phtml.

Schrantz, Michael. "This Land Is His Land: Remembering Bob Armstrong." *Texas Observer*, April 20, 2019. https://www.texasobserver.org/this-land-is-his-land-remembering-bob-armstrong/.

Sheidlower, Noah. "The Controversial History of Levittown, America's First Suburb." *Untapped New York*, n.d. https://untappedcities.com/2020/07/31/the-controversial-history-of-levittown-americas-first-suburb.

Sellers, Christopher. "How Republicans Came to Embrace Anti-Environmentalism." *Vox*, April 22, 2017. https://www.vox.com/2017/4/22/15377964/republicans-environmentalism.

Siciliano, Rocco C. "The Nixon Pay Board—a Public Administration Disaster." *Public Administration Review* 62, no. 3 (2002): pp. 368–373. https://doi.org/10.1111/1540–6210.00187.

Smith, Carter P. "From the Pen of Carter P. Smith." *Texas Parks & Wildlife*, August 2010, https: //tpwmagazine.com/archive/2011/aug/atissue.

Smith, Forrest. "Five Questions for Andy Sansom." *Texas Observer*, March 25, 2010. https://www.texasobserver.org/five-questions-for-andy-sansom/.

Smith, Griffin. "Forgotten Places. Where Something of the Original Texas Still Survives." *Texas Monthly*, August 1975. https://www.texasmonthly.com/articles/forgotten-places-2/.

Swartz, Mimi. "Who Is Gregg Abbott?" *Texas Monthly*, May 2022. https://www.texasmonthly.com/news-politics/who-is-greg-abbott/.

"Taxing the ABC Transaction: A Suggested Approach." *University of Pennsylvania Law Review*, vol. 114 (1966), p. 588–613. https://scholarship.law.upenn.edu/cgi/viewcontent.cgi?article=6267&context=penn_law_review.

Thomas, Evan. "The Shot Heard Round the World." *Newsweek*, March 14, 2010. https://www.newsweek.com/shot-heard-round-world-113499.

Tollefson, Jeff. "Climate Change Is Hitting the Planet Faster than Scientists Originally Thought." *Nature*, February 28, 2022. https://www.nature.com/articles/d41586-022-00585-7.

Votteler, Todd H., ed. "An Internet for Water: Connecting Texas Water Data." *Texas Water Journal* 10, no. 1 (2019): pp. 24–31. https://twj-ojs-tdl.tdl.org/twj/article/view/7086/pdf.

Wagner, Travis. "Hazardous Waste: Evolution of a National Environmental Problem." *Cambridge Core*, April 27, 2009. https://www.cambridge.org/core/journals/journal-of-policy-history/article/abs/hazardous-waste-evolution-of-a-national-environmental-problem/19ABDCF870811BC52B00D6A4E25674C4.

Woods, Neal D. "Why Conservatives Abandoned Conservation Science." *Science*, November 9, 2018. https://www.science.org/doi/10.1126/science.aav2324.

Worth, Katie. "Subverting Climate Science in the Classroom." *Scientific American*, July 1, 2022. https://www.scientificamerican.com/article/subverting-climate-science-in-the-classroom/.

Wilder, Forrest. "Five Questions for Andy Sansom." *Texas Observer*, March 25, 2010. https://www.texasobserver.org/five-questions-for-andy-sansom/.

Websites

Government and University Websites

Architectural Archives Oral History Project. "Emily Little." Interview by Toni Thomasson. Austin Public Library, June 2, 2017. https://library.austintexas.gov/ahc/architectural-archives-oral-history-project#Emily%20Little.

Bengsten, Shawn, Randy Blankinship, and Craig Bonds. "Texas Parks & Wildlife Department, History, 1963–2003." Texas Parks and Wildlife Department. 2003. https://tpwd.texas.gov/publications/pwdpubs/media/pwd_rp_e0100_1144.pdf.

Boccaletti, Guilio. "Water for People or Nature Is a False Choice. We Need to Think Bigger to Protect the World's Water." World Economic Forum, July 11, 2017. https://www.weforum.org/agenda/2017/07/water-for-people-or-nature-is-a-false-choice-we-need-to-think-bigger-to-protect-the-worlds-water/.

Boyd, Jade. "Seven Research Teams Win Carbon Hub Funding," Rice University, *Rice News*, March 8, 2021. https://news.rice.edu/news/2021/seven-research-teams-win-carbon-hub-funding.

Briscoe Center for American History, "Andy Sansom Interview, Part 2 of 2." April 13, 2002. https://digitalcollections.briscoecenter.org/item/3110.

Briscoe Center for American History, "David Bamberger Interview, Part 1 of 2." June 17, 1999. https://digitalcollections.briscoecenter.org/item/2765?solr_nav%5Bid%5D=13321cdc8ec17957611f&solr_nav%5Bpage%5D=0&solr_nav%5Boffset%5D=0

Bush School of Government and Public Service, Texas A&M University. "Texas vs. The Federal Government: An Examination of the Influence of Political Ideologies on State Filed Lawsuits." Spring 2017. https://bush.tamu.edu/wp-content/uploads/2020/02/Spring-2017-Legislative-Capstone-Report.pdf.

Burke, Amy, Abigail Okrent, and Katherine Hale. "Science & Engineering Indicators," National Science Foundation. January 2022. https://ncses.nsf.gov/pubs/nsb20221/conclusion.

Cahn, Robert. "Albright Lecture 1980: The Conservation Challenge of the 80s." UC Berkeley Rausser College of Natural Resources. February 19, 1980. https://nature.berkeley.edu/albright/1980.

Card, David, and Thomas Lemieux, "Did Draft Avoidance Raise College Attendance During the Vietnam War?" Center for Labor Economics, University of California, Berkley. Working Paper No. 46, February 2002. http://cle.berkeley.edu/wp/wp46.pdf.

Chaudhury, Pourab, and Debanjan Banerjee, "'Recovering with Nature': A Review of Ecotherapy and Implications for the COVID-19 Pandemic." US National Library of Medicine, December 10, 2020. https://www.ncbi.nlm.nih.gov/pmc/articles/PMC7758313/.

Connally, Wendy, ed. "Texas Conservation Action Plan 2011—2016: Texas Blackland Prairies" Texas Parks and Wildlife Department. 2011. https://tpwd.texas.gov/landwater/land/tcap/documents/tcap_tbpr_handbook.pdf.

Cough-Schulze, Chantal. "Health at the Nexus of Water Insecurity," Texas Water Resources Institute. Winter 2020. https://twri.tamu.edu/publications/txh2o/2020/winter-2020/health-at-the-nexus-of-water-insecurity/.

Environmental Protection Agency. "What Is Environmental Education?" Accessed July 30, 2022. https://www.epa.gov/education/what-environmental-education.

Environmental Protection Agency. "Why Is Habitat Restoration Near the Gulf of Mexico Essential? Updated November 2, 2023. https://www.epa.gov/gulfofmexico/why-habitat-restoration-near-gulf-mexico-essential.

Fahlquist, Lynne, and R. N. Slattery. "Water-Quality Summary of the San Marcos Springs Riverine System, San Marcos, Texas, July-August 1994." United States Geological Survey. November 2016. https://pubs.usgs.gov/fs/Fs05997/.

Ferguson, Laura. "The Extinction Crisis." Tufts University, *Tufts Now*, May 21, 2019. https://now.tufts.edu/2019/05/21/extinction-crisis.

Harvey, Tom. "Research Documents Child-Nature Disconnect, Shows 'Life's Better Outside.'" Texas Parks Wildlife Department, June 9, 2008. https://tpwd.texas.gov/newsmedia/releases/? req=20080609a.

Harvey, Tom. "TPW Magazine Offers '50 Ways to Get Kids Hooked on the Outdoors.'" Texas Parks Wildlife Department, February 18, 2008. https://tpwd.texas.gov/newsmedia/releases/? req=20080218b.

Henson, Jim, and Joshua Blank, "Texas Politics Project at the University of Texas Austin." May 4, 2022. https://texaspolitics.utexas.edu/blog/new-uttexas-politics-project-poll-texans%E2%80%99-attitudes-population-growth-and-state%E2%80%99s-future-take.

Hess, Myron J. "In Stream Flows—An Environmental Perspective on Texas Laws," Texas A&M University, September 2005. https://texaswater.tamu.edu/readings/ef_plicy/instreamflows.hess.pdf.

Kamarck, Kristy N. "The Selective Service System and Draft Registration: Issues for Congress." Congressional Research Service, Washington, DC. Updated August 18, 2021. https://sgp.fas.org/crs/misc/R44452.pdf.

Kaufman, Andrew. "What's in a Name? An inside Look at Algiers H. Meadows." SMU, *MPRINT Magazine*, 2012. https://blog.smu.edu/meadows50/2019/03/29/whats-in-a-name-an-inside-look-at-algur-h-meadows/.

Knutson, Tom. "Global Warming and Hurricanes." Geophysical Fluid Dynamics Laboratory. October 24, 2022. https://www.gfdl.noaa.gov/global-warming-and-hurricanes/.

Larson, Lincoln R. et al., "Greenspace and Park Use Associated with Less Emotional Distress among College Students in the United States during the COVID-19

Pandemic." US National Library of Medicine. March 2022. https://www.ncbi. nlm.nih.gov/pmc/articles/PMC8648327/.

Longley, W. L. ed. "Freshwater Inflows to Texas Bays and Estuaries: Ecological Relationships and Methods for Determination of Needs." Texas Water Development Board and Texas Parks and Wildlife Department. http://midge-water.twdb.texas.gov/bays_estuaries/Publications/FreshwaterInflows-%20 Ecological%20Relationships%20and%20Methods%20for%20Determintion %20of%20Needs%20-%201994.pdf.

Los Caminos del Rio Heritage Project. "Collection Overview" University Library, Special Collections and University Archives, University of Texas Rio Grande Valley, Edinburg, TX. February 12, 2018. https://archives.lib.utrgv.edu/ repositories/2/resources/257.

Lund, Alison. "Map of the Month: Conservation Easements in Texas." Texas A&M Natural Resources Institute. March 1, 2019. https://nri.tamu.edu/blog/2019/ march/map-of-the-month-conservation-easements-in-texas/.

Maidment, David, et al., "Scientific Principles for Definition of Environmental Flows," University of Texas Civil Architectural and Environmental Engineering. October 31, 2005. https://www.caee.utexas.edu/prof/maidment/nrc/TexasInstream/ EnvironmentalFlowsConference/EnvFlowPrinciples.pdf

McEachern, George Ray. "A Texas Grape and Wine History." Aggie Horticulture, Texas A&M University, 2003. https://aggie-horticulture.tamu.edu/southern-garden/Texaswine.html.

McGrath, Glenn. "Electric Power Sector CO_2 Emissions Drop as Generation Mix Shifts from Coal to Natural Gas," US Energy Information Administration Today in Energy, June 9, 2021. https://www.eia.gov/todayinenergy/detail. php?id=48296.

Meadows Center for Water and the Environment. "Endangered Species." Texas State University. https://www.meadowscenter.txst.edu/ExploreSpringLake/ EndangeredSpecies.html.

Meadows Center for Water and the Environment. "Texas Environmental Flows Initiative Final Report." Texas State University. March 2019. https://gato-docs. its.txstate.edu/jcr:5580662d-e40c-439c-9b76-687f0abc862d.

Montana State Legislature. "Water Policy Interim Committee Dolan—Water Necessary Exempt Wells." Power Point Presentation. https://leg.mt.gov/content/ Committees/Interim/2007_2008/water_policy/staffmemos/typicalneed.pdf.

Nielsen-Gammon, John, Sara Holman, Austin Buley, Savannah Jorgensen, Jacob Escobedo, Catherine Ott, Jeramy Dedrick, and Ali Van Fleet, "Assessment of Historic and Future Trends of Extreme Weather in Texas, 1900–2036." Texas A&M University, Office of the Texas State Climatologist, October 7, 2021. https:// climatexas.tamu.edu/files/ClimateReport-1900to2036-2021Update.

New Mexico Land Conservancy, "Robert B. Anderson." Accessed December 14, 2021. https://nmlandconservancy.org/people/robert-b-anderson.

Office of the Texas Governor. "Governor Abbott Issues Letter to President Biden on EPA Plan That Threatens Oil and Gas Production in Permian Basin." June

27, 2022. https://gov.texas.gov/news/post/governor-abbott-issues-letter-to-president-biden-on-epa-plan-that-threatens-oil-and-gas-production-in-permian-basin.

Peters, Gerhard, and John T. Woolley, "Republican Party Platform of 1980," The American Presidency Project. https://www.presidency.ucsb.edu/node/273420.

Rosen, Rudolph. "H2O Partner, River Systems Institute, to Enhance Outdoor STEM Education Opportunities for Students at Spring Lake, Headwaters of the San Marcos River." H2O Headwaters to Ocean, December 15, 2011. https://www.water-texas.org/river-systems-institute-education-rudy-rosen/.

Rosen, Rudolph. "Meadows Center for Water and the Environment," H2O Headwaters to Ocean, May 1, 2013. https://www.water-texas.org/meadows-center-water-environment-h2o/.

Sansom, Andrew. "A Time in Our Lives." Commencement address presented by Rogers C. B. Morton at the University of Maryland Commencement, January 25, 1971. Rogers C. B. Morton Collection, 1939–76, University of Kentucky Libraries. https://exploreuk.uky.edu/fa/findingaid/?id=xt71g15t7b14.

Smeins, Fred, Sam Fuhlendorf, and Charles Taylor. "Environmental and Land-Use Changes: A Long-Term Perspective." Texas Natural Resource Server. https://texnat.tamu.edu/library/symposia/juniper-ecology-and-management/environmental-and-land-use-changes-a-long-term-perspective/.

Smithsonian's National Zoo and Conservation Biology Institute. "Migratory Bird Center." Accessed June 3, 2021. https://nationalzoo.si.edu/migratory-birds.

Smithsonian's National Zoo and Conservation Biology Institute. "Stopover Habitat along the Gulf of Mexico." May 23, 2019. https://nationalzoo.si.edu/migratory-birds/stopover-habitat-along-gulf-mexico.

Strohl, Sarah. "The Fascinating History of One of the Oldest Places in Texas." Visit San Marcos. March 19, 2018. https://www.visitsanmarcos.com/blog/post/the-fascinating-history-of-one-of-the-oldest-places-in-texas/.

Sunset Advisory Commission. "Executive Summary of Sunset Staff Report: Texas Commission on Environmental Quality; Texas Low-Level Radioactive Waste Disposal Compact Commission. 2022–23. https://www.sunset.texas.gov/public/uploads/2022-05/Texas%20Commission%20on%20Environmental%20Quality%20Staff%20Report_5-25-22.pdf.

Sunset Advisory Commission. "Staff Report: Texas Commission on Environmental Quality; Texas Low-Level Radioactive Waste Disposal Compact Commission." 2022–23. https://www.sunset.texas.gov/public/uploads/2022-05/Texas%20Commission%20on%20Environmental%20Quality%20Executive%20Summary_5-25-22.pdf.

Sunset Advisory Commission. "Texas Water Development Board; State Water Implementation Fund for Texas Advisory Committee." Sunset Staff Report with Commission Decisions. 2022. https://www.sunset.texas.gov/public/uploads/2022-07/Water%20Development%20Board_State%20Water%20Implementation%20Fund%20Staff%20Report%20with%20Commission%20Decisions_7-1-22.pdf.

Sunset Advisory Commission. "Texas Commission on Environmental Quality;
    Texas Low-Level Radioactive Waste Disposal Compact Commission." 2022.
    https://www.sunset.texas.gov/public/uploads/2022-05/Texas%20Commission
    %20on%20Environmental%20Quality%20Staff%20Report_5-25-22.pdf.
Suttie, Jill. "How to Protect Kids from Nature-Deficit Disorder," University of
    California, Berkeley, Greater Good, September 2016. https://greatergood.
    berkeley.edu/article/item/how_to_protect_kids_from_nature_deficit_disorder.
Texas A&M University. "Plants of Texas Rangelands." https://rangeplants.tamu.edu/
    plant/greenbriar/.
Texas A&M Institute of Renewable Natural Resource. "Private Lands Public Benefits."
    2010. https://www.landcan.org/pdfs/Texas-PrivateLandsPublicBenefits.pdf.
Texas Commission on Environmental Quality. "List of Texas PWSs Limiting
    Water Use to Avoid Shortages." https://www.tceq.texas.gov/drinkingwater/trot/
    droughtw.html.
Texas General Land Office and Texas Nature Conservancy. "Natural Heritage of
    Texas." Texas A&M University of Galveston. January 1, 1987. https://tamug-ir.
    tdl.org/handle/1969.3/27376.
Texas Parks and Wildlife Department. "2001 Annual Report." December 1, 2001.
    https://tpwd.texas.gov/publications/pwdpubs/media/pwd_bk_e0100_0003_
    01_02.pdf.
Texas Parks and Wildlife Department. "Big Bend Ranch State Park History." Accessed
    May 24, 2022. https://tpwd.texas.gov/state-parks/big-bend-ranch/park_history.
Texas Parks and Wildlife Department. "Big Bend Ranch Visitor Center Named for
    Bob Armstrong." August 25, 2014. https://tpwd.texas.gov/newsmedia/releases
    /?req=20140825c.
Texas Parks and Wildlife Department. "Devils River State Natural Area History."
    Accessed September 5, 2022. https://tpwd.texas.gov/state-parks/devils-river/
    park_history.
Texas Parks and Wildlife Department. "Devils River State Natural Area Trails Info."
    Accessed July 21, 2022. https://tpwd.texas.gov/state-parks/devils-river/trails-info.
Texas Parks and Wildlife Department. "Eastern Bluebird (Sialia sialis)." https://tpwd.
    texas.gov/huntwild/wild/species/easternbluebird/.
Texas Parks and Wildlife Department. "Financial Overview." December 2020. https://
    tpwd.texas.gov/publications/pwdpubs/media/pwd_rp_a0900_0679_12_20.pdf.
Texas Parks and Wildlife Department. "Mad Island WMA." https://tpwd.texas.gov/
    huntwild/hunt/wma/find_a_wma/list/?id=39.
Texas Parks and Wildlife Department. "Northern Harrier (Circus cyaneus)." https://
    tpwd.texas.gov/huntwild/wild/species/harrier/.
Texas Parks and Wildlife. "Texas Central Coast Region." 2012. tpwd.texas.gov/
    huntwild/wild/wetlands/central-coast/media/central_coast_wmas.pdf.
Texas Parks and Wildlife Department. "Texas Parks and Wildlife Commission Public
    Hearing, August 30, 2001." https://tpwd.texas.gov/business/feedback/meetings/
    2001/0830/transcripts/public_hearing.

Texas Parks and Wildlife Department. "Self Evaluation Report, August 1999." https://
    tpwd.texas.gov/publications/nonpwdpubs/media/tpwd_sunset_self_evaluation
    _report.pdf

Texas Parks and Wildlife Department. "Summary of Minutes Texas Parks and
    Wildlife Commission Conservation Committee, June 2, 1999." https://tpwd.
    texas.gov/business/feedback/meetings/1999/0826/agenda/conservation_
    committee/#header.

Texas Parks and Wildlife Department. "Texas Parks and Wildlife Department
    Involvement in Water Issues." Accessed July 22, 2022. https://tpwd.texas.gov/
    landwater/water/conservation/water_resources/legal/.

Texas Parks and Wildlife Department. "TPWD Postpones Devils River Land
    Acquisition." November 2, 2010. https://tpwd.texas.gov/newsmedia/releases/?
    req=20101102b.

Texas State Library and Archives Commission. "Texas Water Issues: Water in Texas."
    Last Updated May 26, 2026. https://www.tsl.texas.gov/lobbyexhibits/water.

Texas State University. "Department of Geography and Environmental Studies."
    http://mycatalog.txstate.edu/undergraduate/liberal-arts/geography/.

Texas State University. "Former Texas State President Jerome Supple Dies at 67."
    January 16, 2004. https://news.txst.edu/about/news-archive/press-releases/
    2004/01/supple011604.html.

Texas State University. "Meadows Foundation Lends Support to SWT Water
    Resources Effort." November 7, 2002. https://news.txst.edu/about/news-archive/
    press-releases/2002/11/meadowsfoundation110702.html.

Texas State University. "Texas Rivers Center at San Marcos Springs SWT, Parks &
    Wildlife Announce Partnership." October 30, 2000. https: //news.txst.edu/
    about/news-archive/press-releases/2000/10/riverscentpw103000.html.

Texas State University. "Water Policy in Texas: A Comprehensive Overview." https://
    gato-docs.its.txst.edu/jcr:d175e07f-0d03-40bb-b151-4e90b536843a/Water_Policy
    _in_Texas_A_Comprehensive_Overview_2013.pdf.

Texas Water Development Board. "Environmental Flows." Accessed July 30, 2022.
    https://www.twdb.texas.gov/surfacewater/flows/index.asp.

Texas Water Development Board. "The Texas Water Plan." November 1968. https://
    www.twdb.texas.gov/publications/State_Water_Plan/1968/1968_Texas_Water_
    Plan.pdf.

The Meadows Foundation. "Our Critical Water Issues." July 23, 2021. https://www
    .mfi.org/meadows-water-center/.

Thiaw, Ibrahim. "By 2050, 90% of Land Could Become Degraded. How Can
    Businesses Help Restore the Resources They Depend upon?" World Economic
    Forum, 2022. https://www.weforum.org/agenda/2022/01/how-businesses-can-
    help-restore-land-resources/.

United Nations. "Only 11 Years Left to Prevent Irreversible Damage from Climate
    Change, Speakers Warn during General Assembly High-Level Meeting." UN
    Meetings Coverage and Press Releases, March 28, 2019. https://press.un.org/
    en/2019/ga12131.doc.htm.

United Nations. "UN Report: Nature's Dangerous Decline 'Unprecedented'; Species Extinction Rates 'Accelerating.'" Sustainable Development Goals, May 5, 2019. https://www.un.org/sustainabledevelopment/blog/2019/05/nature-decline-unprecedented-report/.

US Commission on Ocean Policy. "Chapter 24: Managing Offshore Energy and Other Mineral Resources," in *An Ocean Blueprint for the 21st Century*. 2004. https://govinfo.library.unt.edu/oceancommission/documents/full_color_rpt/24_chapter24.pdf.

US Department of the Interior. "The Conservation Legacy of Theodore Roosevelt." February 14, 2020. https://www.doi.gov/blog/conservation-legacy-theodore-roosevelt.

US Fish and Wildlife Service. "Service Completes Initial Reviews on Endangered Species Act Petitions for Four Species." October 18, 2022. https://www.fws.gov/press-release/2022-10/service-completes-initial-reviews-endangered-species-act-petitions-four.

Wagner, Aaron. "How Has the COVID-19 Pandemic Affected Outdoor Recreation in America?" Penn State University, January 24, 2022. https://www.psu.edu/news/health-and-human-development/story/how-has-covid-19-pandemic-affected-outdoor-recreation-america/.

Wait, Miranda, and Meagan Lobban, "Flowing a New Path: The Journey in Rebranding Ourselves from Oldest Theme Park in Texas to Successful Nature Center," Texas State University Meadows Center for Water and the Environment. February 21, 2018. https://gato-docs.its.txst.edu/jcr:3be50d3c-510c-43be-9a30-7cbaacaba823.

World Health Organization. "Climate Change and Health." October 30, 2021. https://www.who.int/news-room/fact-sheets/detail/climate-change-and-health.

Yale Program on Climate Change Communication. "Estimated % of Adults Who Think Global Warming Is Happening (Nat'l Avg. 72%), 2021." Accessed February 23, 2022. https://climatecommunication.yale.edu/visualizations-data/ycom-us/.

Yale Program on Climate Change Communication. "Texas Public Opinion on Climate Change 2021." Yale Climate Opinion Maps. Accessed February 23, 2022. https://climatecommunication.yale.edu/visualizations-data/ycom-us/.

Environment websites

Adams, Rod. "Robert O. Anderson—Banking Heir, Oil Wildcatter, Big Oil Exec, Financier of Antinuclear Movement." Nuclear Newswire. August 6, 2013. https://www.ans.org/news/article-1392/robert-anderson-antinuclear-financier.

American Association for the Advancement of Science. "The Reality, Risks, and Response to Climate Change." February 25, 2015. https://whatweknow.aaas.org/get-the-facts/.

Animalia. "White-Winged Dove." https://animalia.bio/white-winged-dove.

BCarbon. "BCarbon Issues First International Soil Carbon Credits in United Kingdom." July 11, 2022. https://static1.squarespace.com/static/611691387b74c566a67f385d/

t/62ccofcc1e6a7f3db3714313/1657540556456/2022-07-11-BCarbon-FFS-Press-Release.pdf.

Berger, Brigid. "Mad Island Smithsonian Banding." Texas Master Naturalist Mid-Coast Chapter. https://midcoast-tmn.org/mad-something.

Billingsley, Seth, Luke Metzger, and Sam Berman. "A Most Valuable Legacy—Investing in the next 100 Years for Texas' State Park System." Environment Texas Research and Policy Center/Public Interest Network. 2022. https://publicinterestnetwork.org/wp-content/uploads/2022/08/FRG-TXE-Parks-Report-Jul22-1.4-web.pdf.

Blackland NPAT. "Prairies of North Texas." Accessed February 17, 2022. https://texasprairie.org/prairies-of-north-texas.

Boynton, Thomas, and Madeline Flynn. "Houston, We Have a Water Problem: Lessons for Urban Water Security." Prevention Web. April 19, 202. https://www.preventionweb.net/news/houston-we-have-water-problem-lessons-urban-water-security.

Calming the Waters. "About Jim." https://www.calmingthewaters.com/about-jim.

Canada Energy Regulator. "Residential and Commercial Landowners." November 21, 2022. http://www.cer-rec.gc.ca/en/safety-environment/damage-prevention/residential-commercial-landowners/index.html.

Carbon Credits. "The Ultimate Guide to Understanding Carbon Credits." March 7, 2022. https://carboncredits.com/the-ultimate-guide-to-understanding-carbon-credits/.

Citizens Preserving Floyd County. "A Landowner's Know Your Rights and Options Manual." August 10, 2014. https://www.preservefloyd.org/wp-content/uploads/2014/08/Know-your-rights-Landowers-Handbook-final.pdf.

Climate Grades. "Making the Grade?" https://climategrades.org/.

Climate Reality Project. "Why Is 1.5 Degrees the Danger Line for Global Warming?" March 18, 2019. https://www.climaterealityproject.org/blog/why-15-degrees-danger-line-global-warming.

Dion, James. "Mad Island Marsh Preserve." Explore Lone Star Coastal, Lone Star Coastal Alliance. December 8, 2020. https://explorelonestarcoastal.com/listing/mad-island-marsh-preserve.

Eckhardt, Gregg. "San Marcos Springs." The Edwards Aquifer website. https://www.edwardsaquifer.net/sanmarcos.html.

Eichler, Sarah, et al. "Abandoned Farmland Restoration." Project Drawdown. https://drawdown.org/solutions/abandoned-farmland-restoration.

Environmental Stewardship. "Environmental Flow Standards." https://www.environmental-stewardship.org/environmental-flows-allocation-process/.

EnvironmentTexas.org. "Million Acre Parks Project." https://publicinterestnetwork.org/wp-content/uploads/2022/09/MAPP-Facts-Sheet-4.pdf.

Hanson, Arthur J. "The First Earth Day: A Founder of the Original Teach-in Remembers." International Institute for Sustainable Development, April 21, 2020. https://www.iisd.org/articles/insight/first-earth-day-founder-original-teach-remembers.

Hill Country Alliance. "Texas Hill Country Conservation Network." https://hillcountryalliance.org/our-work/texas-hill-country-conservation-network.

Hill Country Alliance. "Water Planning." https://hillcountryalliance.org/our-work/water-resources/water-planning/.

Hill Country Land Trust. "Managing Ashe Juniper (Cedar)." https://acrobat.adobe.com/link/review?uri=urn:aaid:scds:US:9d01c912-e863-3784-9e4b-1e89904b096d.

Igini, Martin. "4 Environmental Issues in Texas in 2022." Earth.Org. June 1, 2022. https://earth.org/texas-climate-change/.

Island Conservation. "Mona Island, Restoration Project, Puerto Rico." March 19, 2020. https://www.islandconservation.org/mona-island-puerto-rico/.

Jesperson, Lizzie. "My Living Waters: Dianne Wassenich's Lifelong Mission to Protect the San Marcos River." Texas Living Waters Project. May 1, 2017. https://texaslivingwaters.org/dianne-wassenich-san-marcos-river-foundation/.

Keeler, Dorothy, and Lee Keeler. "A Toklat Wolf Experience Denali National Park and Preserve." Wolf Song of Alaska, n.d., https://wolfsongalaska.org/chorus/A-Toklat-Wolf-Experience-Denali-National-Park.

LandGate Resources. "Carbon Credits Explained: What Are Carbon Credits and How Do They Work?" September 27, 2021. https://landgate.com/news/2021/09/27/what-are-carbon-credits-and-how-do-they-work.

Leffer, Lauren. "The Planet Is Undergoing an Ecological Transformation, Imperiling Biodiversity Everywhere." National Audubon Society. March 4, 2022. https://www.audubon.org/news/the-planet-undergoing-ecological-transformation-imperiling-biodiversity.

Mace, Robert E. "Groundwater and the White Shaman." So secret, Occult, and Concealed: The Story of Groundwater in Texas. March 17, 2020. https://sosecretoccultandconcealed.com/2020/03/07/groundwater-and-the-white-shaman/.

McCormick, Andrew. "65% Of Texans Worry about Climate, but 'Broadscale Voter Suppression' Impedes Action." The Energy Mix. April 28, 2022. https://www.theenergymix.com/2022/04/28/65-of-texans-worry-about-climate-but-broadscale-voter-suppression-impedes-action/.

Metzger, Luke, and Susan Kaplan, "Actor Ethan Hawke Backs Texas State Parks Expansion Push." Environment Texas. August 25, 2022. https://environmentamerica.org/texas/updates/actor-ethan-hawke-backs-texas-state-parks-expansion-push/.

Montgomery, David. "Growing A Revolution: Excerpt." Resilience. June 22, 2022. https://www.resilience.org/stories/2022-06-22/growing-a-revolution-excerpt/.

Montgomery, David R. "The Novelist Who Loved Soil." Nature News. April 14, 2020. https://www.nature.com/articles/d41586-020-01024-1.

More, Thomas A. "From Public to Private: Five Concepts of Park Management and Their Consequences." The George Wright Forum. 2005. www.georgewright.org/222more.pdf.

Net Zero Climate. "What Is Net Zero?" August 23, 2022. https://netzeroclimate.org/what-is-net-zero.

National Audubon Society. "History of Audubon and Science-Based Bird Conservation." June 4, 2018. https://www.audubon.org/about/history-audubon-and-waterbird-conservation.

National Environment Education Foundation. "Benefits of Environmental Education." 2020. https://www.neefusa.org/education/benefits.

Net Zero Climate. "What Is Net Zero?" Accessed August 23, 2022. https://netzeroclimate.org/what-is-net-zero/.

Noonan, Patrick. "An Interview with Our Founder." The Conservation Fund. 2015. https://www.conservationfund.org/30th-anniversary/an-interview-with-our-founder.

Oldmixon, Seth. "Future Trends of Extreme Weather in Texas." Texas 2036. October 12, 2021. https://texas2036.org/weather/.

Quivira Coalition. "Mollie Walton." https://quiviracoalition.org/mollie-walton/.

Randall, Brianna. "The Stream Whisperer: 'Thinking like Water' Restores Sage Grouse Habitat," Cool Green Science. August 23, 2017. https://blog.nature.org/science/2017/08/23/stream-whisperer-thinking-like-water-restores-sage-grouse-habitat/.

Sage Grouse Initiative. "Starter Guide for Healing Degraded Meadows with Hand-Built Structures in Sagebrush Country." June 4, 2018. https://www.sage-grouseinitiative.com/starter-guide-for-healing-incised-meadows-with-hand-built-structures-in-sagebrush-country/.

San Marcos River Foundation. "Dianne Wassenich." https://sanmarcosriver.org/award-recipient/dianne-wassenich.

Save Buffalo Bayou. "Frank Smith, Conservationist." October 16, 2016. www.savebuffalobayou.org/?page_id=3189.

Selah, Bamberger Ranch Preserve. "Our Story." https://www.bambergerranch.org/our-story.

Sierra Club. "Get the Frack out of Our Parks." February 21, 2017. https://www.sierraclub.org/texas/blog/2017/02/get-frack-out-our-parks.

Sierra Club. "Sierra Club Lawsuit Challenges Construction of Permian Highway Fracked Gas Pipeline." April 30, 2020. https://www.sierraclub.org/texas/blog/2020/04/sierra-club-lawsuit-challenges-construction-permian-highway-fracked-gas-pipeline.

Sierra Club. "State Water Planning." https://www.sierraclub.org/texas/state-water-planning.

Stanley, Jim. "The Most Common Grasses of the Hill Country." Hill Country Naturalist.http://hillcountrynaturalist.org/pdf/160618%20The%20Most%20Common%2Grasses%20of%20the%20Hill%20Country.pdf.

Texas Land Trust Council. "Conservation Lands Inventory." Accessed November 2, 2022. https://texaslandtrustcouncil.org/what-we-do/conservation-lands-inventory/.

Texas Land Trust Council. "What Is a Conservation Easement?" Accessed September 23, 2021. https://texaslandtrustcouncil.org/about/what-is-a-conservation-easement/.

Texas Living Waters Project. "The SB3 Environmental Flows Process." https://texaslivingwaters.org/environmental-flows/sb3-environmental-flows-process/.

The Nature Conservancy. "Mad Island Marsh Preserve." https://www.nature.org/en-us/get-involved/how-to-help/places-we-protect/clive-runnells-family-mad-island-marsh-preserve/.

The Nature Conservancy. "Big Bend National Park." December 19, 2019. https://www.nature.org/en-us/about-us/where-we-work/united-states/texas/stories-in-texas/texas-by-nature-big-bend-national-park/.

The Nature Conservancy. "Our Mission, Vision, and Values." https://www.nature.org/en-us/about-us/who-we-are/our-mission-vision-and-values.

The Nature Conservancy. "Texas Water Program." November 5, 2020. https://www.nature.org/en-us/about-us/where-we-work/united-states/texas/stories-in-texas/tx-water-program/.

The Nature Conservancy. "The Water and Nature Declaration." https://www.nature.org/en-us/what-we-do/our-insights/perspectives/nature-for-water-security/.

Van Ballenberghe, Victor. "Savage River Wolf Pack Denali National Park and Preserve." Wolf Song of Alaska. Accessed March 13, 2021. https://wolfsongalaska.org/chorus/Savage-River-Wolf%20Pack-Denali-National-Park-Preserve.

Walls, Margaret A. "The Land and Water Conservation Fund 101." Resources for the Future. https://www.rff.org/publications/explainers/land-and-water-conservation-fund-101/.

Wildlife Habitat Federation. "Wildlife Habitat Federation Overview." https://wildlifehabitatfederation.org/.

World Birding Center. "World Birding Center: 'Nature Adventures in Texas.'" http://www.theworldbirdingcenter.com/.

Other websites

Alaska.org. "Float Trips: Kanektok River." https://www.alaska.org/detail/kanektok-river.

Alden B. Dow Home and Studio. "Tours/Events." https://www.abdow.org/.

American Academy for Park and Recreation Administration, "Terese 'Terry' Tarlton Hershey." 2003. https://aapra.org/Awards/Pugsley-Medal/Recipient-Biography/Id/32.

Angus, Harr. "Mona Island, Puerto Rico." Planeta.com. August 7, 1998. https://www.planeta.com/mona-island-1996/.

Arnold III, J. Barto. "Fort Esperanza." Handbook of Texas Online. Texas State Historical Association. January 1, 1995. https://www.tshaonline.org/handbook/entries/fort-esperanza.

Austin College. "Private Liberal Arts College in Sherman, TX." https://www.austincollege.edu/.

Blakemore, Erin. "How Nixon Became the Unlikely Champion of the Endangered Species Act." History.com/A&E Television Networks. August 19, 2019. https://www.history.com/news/richard-nixon-endangered-species-act-esa-environment.

Britannica. "Basilica of Guadalupe." September 8, 2017. https://www.britannica.com/topic/Basilica-of-Guadalupe.

Blackburn, Jim. "Coastal Living—A Look at a Lifetime of Practicing Environmental Law." State Bar of Texas. https://www.texasbar.com/AM/Template.cfm?Section=articles&Template=%2FCM%2FHTMLDisplay.cfm&ContetID=41508.

Calhoun County Museum. "Matagorda Island." https://calhouncountymuseum.org/exhibits/matagorda-island/.

Cariou, Gerry. "The Nature of Nature Deficit Disorder." Sunset Country. January 18, 2022, https://visitsunsetcountry.com/the-nature-of-nature-deficit-disorder.

Camp Fawcett. "History of Camps Fawcett." https://campfawcett.org/history-of-camp-fawcett.

Chen, Grace. "Climate Change to Become Part of Core Curriculum in Public Schools." Public School Review. May 19, 2022. https://www.publicschoolreview.com/blog/climate-change-to-become-part-of-core-curriculum-in-public-schools.

Clements Texas Papers. "Memo from Paul T. Wrotenbery to William P. Clements regarding Revised Position Statement on Big Bend Ranch Proposal, April 9, 1981." Clements Papers Project. https://tx.clementspapers.org/clementstx/40226.

Cohen, Danielle. "Why Kids Need to Spend Time in Nature." Child Mind Institute. September 21, 2021. https://childmind.org/article/why-kids-need-to-spend-time-in-nature/.

Conservapedia. "J. E. 'Buster' Brown." January 3, 2022. https://www.conservapedia.com/J._E._%22Buster%22_Brown.

Cotswold Natural Mindfulness and Forest Bathing. "The Origin of Forest Bathing and Forest Therapy." March 10, 2019. https://www.ianbanyard.com/home/the-origin-of-forest-bathing-forest-therapy/.

Cox, Mike. "The Naming of Devils River," Texas Escapes, September 2012. http://www.texasescapes.com/MikeCoxTexasTales/Naming-of-Devils-River.htm.

Curlee, Kendall. "Meadows, Algur Hurtle." *Handbook of Texas Online*. Texas State Historical Association. Last Updated June 18, 2021. https://www.tshaonline.org/handbook/entries/meadows-algur-hurtle.

Darboe, Fatou. "Bill Gates and Steve Jobs Limited Screen Time for Their Children." CRM.org. March 15, 2018. https://crm.org/articles/bill-gates-kids-and-steve-jobs-limited-screen-time.

Deloitte United Kingdom. "Working towards Net-Zero Carbon Emissions by 2030." https://www2.deloitte.com/uk/en/pages/annual-report-2020/stories/working-towards-net-zero-emissions-by-2030.html.

Donell Kohout, Martin. "Big Bend Ranch State Park." Handbook of Texas Online. Texas State Historical Association. November 1, 1994. https://www.tshaonline.org/handbook/entries/big-bend-ranch-state-park.

Dow Chemical Company. "Dow Texas Operations." https://corporate.dow.com/en-us/locations/freeport.html.

Earth Day. "History of Earth Day." May 11, 2022. https://www.earthday.org/history/.

EduRank. "10 Best colleges for Hydrology and Water resources management in Texas." May 2022. https://edurank.org/environmental-science/hydrology/texas/.

Encyclopedia.com. "Trends in the Environmental Sciences since 1950." Accessed November 30, 2022. https://www.encyclopedia.com/science/encyclopedias-almanacs-transcripts-and-maps/trends-environmental-sciences-1950.

ExploreNorth. "The History of Quinhagak." https://explorenorth.com/library/communities/alaska/bl-Quinhagak.htm.

Gass, Joyce M. "Honey Creek, TX." Handbook of Texas Online. Texas State Historical Association. Updated February 1, 1995. https://www.tshaonline.org/handbook/entries/honey-creek-tx.

Getchell, Michelle. "The Student Movement and the Antiwar Movement." Khan Academy. https://www.khanacademy.org/humanities/us-history/postwarera/1960s-america/a/the-student-movement-and-the-antiwar-movement.

Great Nonprofits. "2016 Top-Rated Awards: States with the Most Nonprofits." November 21, 2016. https://blog.greatnonprofits.org/2016-top-rated-awards-states-with-the-most-nonprofits/.

Gulf Coast Cattleman. "Coalition of Property Rights Advocates Applaud Latest Eminent Domain Legislation." http://www.gulfcoastcattleman.com/coalition-of-property-rights-advocates-applaud-latest-eminent-domain-legislation.

Hanson, John R. II. "Review of Scarcity and Frontiers: How Economies Have Developed through Natural Resource Exploitation, by Edward B. Barbier." August 2011. https://eh.net/book_reviews/scarcity-and-frontiers-how-economies-have-developed-through-natural-resource-exploitation/.

Harrigan, Stephen. "Comanche Springs." Pecos County Historical Commission. https://www.pecoscountyhistoricalcommission.org/new-page-2.

Hays, Jeffrey. "Singapore's Crazy Rules and Justification for Them." Facts and Details. June 2015. http://factsanddetails.com/southeast-asia/Singapore/sub5_7c/entry-3764.html

History.com. "Nuclear Disaster at Three Mile Island," A&E Television Networks. November 24, 2009. https://www.history.com/this-day-in-history/nuclear-accident-at-three-mile-island.

History.com. "The First Earth Day." A&E Television Networks. November 24, 2009. https://www.history.com/this-day-in-history/the-first-earth-day.

Hulse, James. "San Marcos Mill Tract." The Historical Marker Database. October 6, 2020. https://www.hmdb.org/m.asp? m=157431.

Jones, Mary Beth. "Peach Point Plantation." Handbook of Texas Online. Texas State Historical Association. 2020, www.tshaonline.org/handbook/entries/peach-point-plantation.

Joseph J. Earthman Generations Funeral Home. "In Memory of Terese T. 'Terry' Hershey 1923–2017." Obituary. Accessed February 15, 2021. https://josephjearthman.funeraltechweb.com/tribute/details/147/Terese-Hershey/obituary.html

Kleiner, Diana J. "General American Oil Company." Handbook of Texas Online. Texas State Historical Association. September 1, 1995. https://www.tshaonline. org/handbook/entries/general-american-oil-company.

Lake Texoma Lodge and Resort. "Lake Texoma Lodge and Resort." https://www .laketexomalodge.com/.

Leatherwood, Art. "Matagorda Island." Handbook of Texas Online. Texas State Historical Association. Updated April 1, 1995. www.tshaonline.org/handbook/ entries/matagorda-island.

Levinson, Eve. "The Progressive Party: Ideas and Beliefs—Lesson Transcript." Study. com. Updated November 21, 2023. https://study.com/academy/lesson/the-progressive-party-ideas-beliefs.html.

Louv, Richard. "What Is Nature-Deficit Disorder?" Richard Louv Blog. October 15, 2016. https://richardlouv.com/blog/what-is-nature-deficit-disorder/.

MacLean, A. A. "History of Lake Jackson." City of Lake Jackson. https://lakejackson -tx.gov/267/History.

McAllen Ranch. "McAllen Ranch: A History of Quality Cattle and Horses." Accessed July 7, 2020. mcallenranch.net/history.

McAllen, Margaret H., and Mary Margaret McAllen. "McAllen Ranch." Handbook of Texas Online. Texas State Historical Association. Updated July 18, 2020. www .tshaonline.org/handbook/entries/mcallen-ranch.

Moody, Rebecca. "Screen Time Statistics: Average in the US vs. Rest of the World." Comparitech. March 21, 2022. https://www.comparitech.com/tv-streaming/ screen-time-statistics/.

Nasti, Cecilia. "Centennial Artist Clemente Guzman. Passport to Texas. October 31, 2019. https://passporttotexas.org/category/shows/state-parks/page/2/.

National Mississippi River Museum and Aquarium. "Achievement Award Winners." https://www.rivermuseum.com/national-achievement-award-winners/winners/ jw-jake-hershey.

National Recreation and Park Association. "National Recreation and Park Association." https://www.nrpa.org/.

Nichols, Taylor. "Data Show Today's Environmental Science Grads Have More Diverse Job Prospects." OnlineU. 2022. https://www.onlineu.com/magazine/ environmental-science-majors-career-outcomes.

Nickels, David L., and Britt C. Bousman. "Archaeological Testing at San Marcos Springs (41HY160) for the Texas Rivers Center, Hays County, Texas." Index of Texas Archaeology: Open Access Gray Literature from the Lone Star State: Vol. 2010, Article 8. https://doi.org/10.21112/ita.2010.1.8

Phillips, Bob. "Aquarena Springs." http://www.aquarenaandralph.com/.

Pohl, Kelly, and Megan Lawson. "State Funding Programs for Outdoor Recreation: Texas Sporting Goods." Outdoor Industry Association. https:// headwaterseconomics.org/wp-content/uploads/state-rec-TX.pdf.

Public Citizen. "South Texas Nuclear Project—the Record." https://www.citizen.org/ wp-content/uploads/migration/stnp_chronology.pdf.

Raj Joshi, Sanjaya. "Comparison of Groundwater Rights in the United States: Lessons for Texas." Master's Thesis, Texas Tech University College of Engineering. 2005. https://aquadoc.typepad.com/files/gw_rights_thesis.pdf.

Richards, Adam. "The Student Movement of the 1960s." Study.com. https://study .com/academy/lesson/the-student-movement-of-the-1960s.html.

Real Yellow Pages. "The Brazosport Facts." https://www.yellowpages.com/clute-tx/ mip/the-facts-1435636.

Remember Singapore. "Gongs, Long Hair and Chewing Gums." March 21, 2016. https://remembersingapore.org/2014/06/19/gongs-long-hair-chewing-gums/

Robinson, Eric. "The Challenge of Hunting Deer on Public Land in Texas." The Swift Lift. September 30, 2019. https://www.theswiftlift.com/challenge-hunting-deer-public-land-texas/.

Rothwell, Susan L. "Antinuclear Movement." Britannica. Updated December 28, 2023. https://www.britannica.com/topic/anti-nuclear-movement.

Rowland-Shea, Jenny, and Elyssa Spitzer. "The Most Anti-Nature President in U. S. History." Center for American Progress. May 21, 2020. https://www .americanprogress.org/article/anti-nature-president-u-s-history/.

Sechrist, Peggy. "Hershey Ranch." ArcGIS Story Maps. May 12, 2021. https:// storymaps.arcgis.com/stories/cbc939743729488bb2156de6cc03ebfa.

Shiner, Joel L. "Spring Lake Site." Handbook of Texas Online. Texas State Historical Association. December 1, 1995. https://www.tshaonline.org/handbook/entries/ spring-lake-site.

Smith Fay, Mary. "Fay, Albert Bel (1913–1992)." Handbook of Texas Online. Texas State Historical Association. March 17, 2017. https://www.tshaonline.org/ handbook/entries/fay-albert-bel.

Smyrl, Vivian Elizabeth. "Guadalupe River." Texas State Historical Association. Handbook of Texas Online. Updated October 6, 2022. https://www.tshaonline. org/handbook/entries/guadalupe-river.

Stranahan, Pam. "When Were Towns on Matagorda Island?" History Center for Aransas County. www.theachistorycenter.com/history-mystery-1/when-were-towns-on-matagorda-island%3F.

Tedesco, John. "Forced off Their Land; Joe Hawes and His Family Lost Their Ranch on Matagorda Island to the U. S. Government Many Long Years Ago. They Still Want It Back." John Tedesco.net. April 30, 2001. https://johntedesco. net/blog/forced-off-their-land-joe-hawes-and-his-family-lost-their-ranch-on-matagorda-island-to-the-us-goverment-many-long-years-ago-they-still-want-it-back/.

Texas Freedom Network. "National Report Card: Texas among the States Failing Students on Climate Change." October 8, 2020. https://tfn.org/national-report-card-texas-among-the-states-failing-students-on-climate-change/.

Texas Legacy Project. "Jim Blackburn." Interview by David Todd. October 1, 1999. https://www.texaslegacy.org/transcript/jim-blackburn/.

Texas Legacy Project. "Mickey Burleson." Interview by David Todd. June 19, 1999. https://www.texaslegacy.org/transcript/mickey-burleson/.

Texas Legacy Project, "Terry Hershey." Interview by David Todd. April 13, 2002. https://www.texaslegacy.org/transcript/terry-hershey/.

Texas Legacy Project. "Terry O'Rourke." Interview by David Todd. October 2, 1999. https://www.texaslegacy.org/transcript/terry-orourke/.

Texas Legacy Project. "Sharron Stewart." Interview by David Todd and David Weisman. October 23, 2003. https://www.texaslegacy.org/transcript/sharron-stewart/.

Texas Time Travel. "Clymer Meadow Preserve." https://texastimetravel.com/directory/clymer-meadow-preserve/.

Texas Tribune Festival. "Andy Sansom's Schedule for the Texas Tribune Festival 2012." https://thetexastribunefestival2012.sched.com/speaker/andysansomexecutive directorriversystemsinstituteformerexecutivedirectortexaspa3.

Untermeyer, Chase. "Chase Untemeyer: Full Bio." https://www.untermeyer.com/full-bio/.

USLegal Inc. "ABC Transaction [Oil & Gas] Law and Legal Definition." https://definitions.uslegal.com/a/abc-transaction-oil-gas/.

Weber, Kelsey. "Dropping Technology and Returning to the Great Outdoors." Youth First. March 20, 2018. https://youthfirstinc.org/dropping-technology-and-return-to-the-great-outdoors/.

Whelan, Carly. "Rosillo Peak." Texas Beyond History. www.texasbeyondhistory.net/trans-p/images/ap10.html.

Index

Other titles in the Kathie and Ed Cox Jr. Books on Conservation Leadership, sponsored by The Meadows Center for Water and the Environment, Texas State University

PUBLIC LANDS

National Parks, Preserves, and Wildlife Refuges — 08

State Parks and Historic Sites — 31

State Wildlife Management Areas — 43

State Natural Areas and Coastal Preserves — 08

State Fish Hatcheries and Laboratories — 05

Regional, County, City, and Academic Properties — 05

PRIVATE LANDS

The Nature Conservancy — 02

Other Nonprofits — 02

This map shows parks, wildlife management areas, and other protected lands that Sansom helped create by identifying vulnerable plants, animals, and ecosystems; then negotiating to acquire them for conservation by various public and private entities.

Details for numbered map locations are provided in the Protected Lands Appendix.

Data Sources: Texas Parks & Wildlife Department, The Nature Conservancy, websites for miscellaneous protected areas.

Note: Transactions totaling less than 5 acres are not shown.

99

75
74

17 40

80

Pecos River

13

52

85

15 10

05 46